m327-1
4

Lothar Sachs

A Guide to
Statistical Methods

and to the Pertinent Literature

Literatur zur
Angewandten Statistik

Springer-Verlag
Berlin Heidelberg New York
London Paris Tokyo

Priv.-Doz. Dr. rer. nat. Lothar Sachs
Abteilung Medizinische Statistik
und Dokumentation
Brunswiker Str. 2 A
2300 Kiel 1, FRG

ISBN 3-540-16835-4
Springer-Verlag Berlin Heidelberg New York
ISBN 0-387-16835-4
Springer-Verlag New York Berlin Heidelberg

Library of Congress Cataloging in Publication Data
Sachs, Lothar.
A guide to statistical methods and to the pertinent literature = Literatur zur
angewandten Statistik.
English and German.
Bibliography: p.
1. Mathematical statistics – Bibliography. 2. Statistics – Bibliography.
I. Title. II. Title: Literatur zur Angewandten Statistik.
Z6654.M33S23 1986 016.5195 86-20304 [QA276]
ISBN 0-387-16835-4 (U.S.)

© Springer-Verlag Berlin Heidelberg 1986
Printed in Germany

Typesetting, printing and bookbinding: Appl, Wemding
2141/3140-543210

Contents/Inhalt

1. Some References to Statistical Methods: Introduction . VII

2. Literatur zur Angewandten Statistik: Vorwort und
   Einführung X

3. Terms/Begriffe 1

4. List of Abbreviations of Journal Titles/Abkürzungen
   einiger Zeitschriften-Titel 130

5. References/Literatur 132

6. Software: Some Addresses/Einige Anschriften 209

1. Some References to Statistical Methods: Introduction

Readers of my books, students and scientists, often ask for special references not commonly found in introductory or intermediate books on statistics. From the titles and contents of 1449 key papers and books which are listed and numbered in Section 5, I have selected keywords and subject headings and arranged them alphabetically together with the numbers of pertinent references in Section 3.

Number 1153, for instance, denotes my book "Applied Statistics". It contains a bibliographical section on pages 568 to 641. Supplementary material is displayed in this small bibliographical guide. It also complements well-known textbooks of Box, Hunter and Hunter (No. 121), Dixon and Massey (No. 286), Snedecor and Cochran (No. 1238), and many recent competitors.

Since the methodology of statistics is expanding rapidly, many methods are not considered at all or only introduced in the basic textbooks of statistics. There is a need for intermediate statistical methods concerned with increasingly complicated applications of statistics to actual research situations. Here the specification of terms helps to find some sources. Since the references vary considerably in length and content, the number of culled or extracted terms per referenced page varies even more, as does also their degree of specialization; however in most cases an intermediate statistical level is maintained.

This guide to supplemental methods is intended for students and researchers from different fields who, having some knowledge of statistics, need special hints on familiar intermediate and multivariate methods and on quite recent developments. Much more may be found in the "Encyclopedia of Statistical Sciences" (No. 701) and especially in "The Annual Index to Sta-

tistical Literature" (e.g. No. 403). Moreover, xerocopies or microfilms of pertinent dissertations are available (No. 1331).

For new literature on software packages one can also use the addresses given on page 209. For new statistical algorithms see, for instance, the regular section in the journal "Applied Statistics" (e.g. No. 471 and especially No. 453).

This guide offers about 5500 terms, a lot of them expanded, on a variety of statistical methods with special emphasis on: thinking in models, planning experiments and surveys, applying graphical, data-analytic and multivariate procedures, and using the proper statistical computing technique. References and terms concern statistical methodology for both professional statisticians and nonprofessional users of statistics with different levels of statistical interest and sophistication. The book will be useful for consulting purposes and may help to support statistical expert systems, though here individual experience counts more than knowledge found in literature. Since the numbers of the references following the terms are nearly always a small selection, this is underlined by the insertion of "e.g." (exempli gratia) "for instance" or by the German equivalent "z.B." (zum Beispiel). The current terminology may be partly unknown to the potential user of this book, thus I have tried to guide him/her by means of non-technical expressions, some perhaps rather trivial or odd to the expert. Moreover, I did not standardize the English terms. So you may find "chi-square" and "chi-squared" as I found them. In assembling this material, three more or less subjective selection criteria were applied: (1) the percentages of publications attributed to original papers, reviews and monographs, (2) the actual selection of the listed references and (3) the selection, structuring and wording of the listed terms, largely dependent on the referenced source, most of which are written in English or have at least an English summary, and nearly all of them are easily accessible. For the German reader not so well acquainted with the English language I included some books and papers, some of which concern the demography of the Federal Republic of Germany.

For several years I hesitated to write this book, hoping colleagues with better resources would take the trouble. Finally I attempted it after all. Since the remainder of my workload remained undiminished, it took some time to assemble, structure and unify the material. The reader may be the judge of the appropriateness of my inevitably subjective selection of publications and terms and of the usefulness of this venture.

2. Literatur zur Angewandten Statistik: Vorwort und Einführung

Leser meiner Bücher haben mich seit über 18 Jahren um Literaturhinweise gebeten. Die vorliegende Sammlung enthält häufig erfragte Begriffe mit den jeweiligen Publikationen, in denen diese Begriffe näher erläutert und angewandt werden. Es sind (1) Stichwörter, Keywords des Titels, der Zwischentitel oder Abschnittsüberschriften und (2) frei gebildete den Inhalt erschließende Schlagwörter und Subject Headings. Bei zitierten Büchern sorgen Inhalts- und Sachverzeichnisse für das Auffinden der gesuchten Information. Dem "Terms/Begriffe" genannten Stichwort-Schlagwort-Verzeichnis (SSV) folgen die hauptsächlich aus Originalmitteilungen, Übersichten, Monographien und wenigen Lehrbüchern bestehenden "References/Literatur", deren Juwel die Nummer 701 aufweist. Eine weitere Fundgrube ist der jährlich erscheinende "Index to Statistical Literature" (z. B. Nr. 403). Bescheidene bibliographische Hinweise enthalten auch meine Bücher, Nr. 1152 [S. 437 bis 495] und Nr. 1153 [S. 568 bis 641], die hierdurch ergänzt werden, ebenso wie andere wohlbekannte Lehrbücher (z. B. die Nummern 121, 286 und 1238). In dieser kleinen Bibliographie konnte ich auch Anregungen von Lesern und Rezensenten berücksichtigen. Das SSV ist zweisprachig angelegt (vgl. die "Introduction") und natürlich unvollständig und subjektiv. Der Fachmann wird manches vermissen. Einige Begriffe werden ihm sonderbar erscheinen, anderes mag er belächeln. Seltsame Wortketten und terminologische Extravaganzen dienen dem Leser, der seinen Weg sucht und häufig als "Einsteiger" die Terminologie nur unvollständig beherrscht. Die im SSV angegebenen Nummern bilden nur eine Auswahl. Daher wird diesen Nummern auch häufig "z. B." oder "e. g." vorangestellt. Ursprünglich auf 1000 Literaturangaben geplant,

wuchs die Bibliographie auf 1449 Einheiten. Sie ist subjektiv ausgewählt mit dem Ziel, dem Anwender einen Weg zu weiteren Methoden der Angewandten Statistik aufzuzeigen. Im Vordergrund stehen das Denken in Modellen, die Planung von Experimenten und Erhebungen, graphische, datenanalytische und multivariate Methoden sowie "Statistical Computing".

Für die Bundesrepublik und die USA sind auch einige Datensammlungen angegeben. Umfangreiche international vergleichbare Daten sind den einschlägigen United Nations Publications, New York 10017 bzw. Genf 10, zu entnehmen.

Der Einsteiger wird zunächst eher zitierte Lehrbücher und ältere Originalien konsultieren, der Fortgeschrittene eher neuere Originalien und Übersichten, wobei die den Aufsatz bringende Zeitschrift erste Rückschlüsse auf den anvisierten Leserkreis zuläßt. Eine begrenzte Aktualisierung dieser kleinen Bibliographie ist möglich, wenn neuere Jahrgänge interessierender Zeitschriften und die Kataloge einiger Verlage durchgesehen werden. Diese Studie basiert auf meiner Kartei, die ohne die Bereitschaft unserer Bibliothekarin Frau Katrin Anger, mir viel Schreibarbeit abzunehmen, nicht möglich gewesen wäre. Ihr sei herzlich gedankt. Den Mitarbeiterinnen und Mitarbeitern des Springer-Verlages danke ich für die sehr erfreuliche Zusammenarbeit.

Klausdorf, im Mai 1986 Lothar Sachs

3. Terms/Begriffe

The following list of terms, mostly English, is alphabetically arranged. Few exceptions from this order help to avoid oddities. I have tried to balance rather individual terms of authors with the standardized terminology used in leading books. The former, though sometimes strange, are easier to locate in the cited key publications. Here contents and subject index as well as subheadings help. The terms are followed by often only one or few numbered references of the bibliography on page 132, a naturally not exhaustive selection. The reader interested in German terms and publications may find some interesting hints and an access to the official statistics of the Bundesrepublik Deutschland.

Die meist englischen Begriffe sind alphabetisch angeordnet. Ihnen folgen Nummern, die auf die entsprechenden Literaturangaben der Bibliographie auf S. 132 verweisen. Die deutschen Begriffe dienen dem in der englischen Sprache weniger geübten Leser als Einstieg. Zugleich betrifft ein Teil von ihnen bevölkerungsstatistische Themen und andere Bereiche der amtlichen Statistik der Bundesrepublik Deutschland.

A-B-A Designs in experimental single case research 220, 527

Abbildung der Realität in die Datenbank 104, 576, 1330, 1349

ABDO, abbreviated Doolittle algorithm for estimating regression models, e.g. 15

Abraham's sacrifice, a game theoretic approach 128

Abweichungen von der Gleichverteilung: Endziffern, z. B. 1074

Accelerated:
failure times, e.g. 560
life model, e.g. 234

Acceptance sampling in quality control 700, 781, 1008, 1181

Accident liability and accident proneness, e.g. 1430

Accidents: two parties involved 1394

Accident theory, e.g. 1430

Accumulation and integration of the research results in a special domain: see meta-analytic procedures

Accuracy of measurements 279, 441, 733, 1081, 1387

ACOVS model, a submodel of the LISREL model, e.g. 628

Acquisitions process, reliability in the 265

Actuarial:
estimator and Greenwood's formula, e.g. 234
method (life table), e.g. 194
methods 64, 431, 528, 575, 758, 1300

Adaptive:
filtering, e.g. 4, 811
statistical techniques, a review of some 559
Statistik 157, 559, 879
verteilungsfreie Tests 157, 559, 879

ADDALS algorithm, e.g. 1436

Additivität der Faktorwirkungen, z.B. 1122

Additivity analysis, e.g. 1436

Adjusted rates, e.g. 194, 380

Adjustment:
in life table studies 194
linear, in premeasure/postmeasure design 21
for confounding 1182
for qualitative effects: prediction methods (generalized linear models) 751
in case control studies, e.g. 258
of rates, e.g. 194, 380
versus matching, e.g. 21, 205

Adjustments
in analysis of observational studies 21, 194, 205
of rates: adjusted rates with their standard errors, e.g. 194

Admissibility, e.g. 774

Ähnlichkeitsdaten: siehe mehrdimensionale Skalierung

Ähnlichkeit von Individuen derselben Familie 1138

Ältere Menschen in der Bundesrepublik: Basisdaten u.a.m. 161, 539, 1148

Age:
-adjustment of rates, e.g. 602
-dependent survival and capture rates for capture-recapture models 1064
in cohort analysis, e.g. 851

Ageing, demographic, of Britain and of the USA 818

Age:
of onset distributions, e.g. 332
period
and cohort effects in demography 851
and cohort effects in earnings equations, using longitudinal data to estimate 851
and cohort effects in marital fertility 851
-period-cohort analysis and the study of deaths from pulmonary tuberculosis 851
-sex structure and migration 1281
specific measures of mortality e.g. 194
-specific:
mortality index 602
relative risks 1292

Aging:
and chronic diseases, e.g. 332
the demography of 818, 851, 1148
the sociology of: an annotated bibliography and sourcebook 493

Agrarstatistik 161

Agreement:
among two sets of rankings of k objects 790
and association: see association between rankings 433, 434

Agricultural:
experimentation in a developing country, e.g. 35

field experiments 1036
sampling theory 1159
AID, Automatic interaction detection, e.g. 513
Air/land warfare models 581
Akaike's information criteria, e.g. 890
Alcohol consumption and cigarette smoking 1091, 1429
Algorithmen der Statistik: siehe Kleinrechner, Programmsysteme sowie statistical Computing
Algorithm to obtain critical values of the t, χ^2 and F distributions 191, 841
Algorithms, applied statistical, e.g. 453
Aligned observations, tests for, e.g. 836
Allgemeines lineares Modell: siehe lineare Modelle sowie verallgemeinerte lineare Modelle
Allometry, the study of shape and size, e.g. 422, 840
ALSOS programs, e.g. 1436
Altengeneration und Verschiebung der Alterspyramide 161, 539, 1148, 1415
Alten-Statistik: ältere Menschen 161, 340, 539, 818, 851, 1148, 1255, 1415
Altern: Bevölkerungsstatistik 340, 818, 851, 1148, 1255
Altersstandardisierung in Bevölkerungsstatistik und Epidemiologie 357, 367, 602
Altersstruktur, z. B. 340, 505
Altersversicherung 161
Alter und Sterblichkeit 161, 1255
Alterungsindex und Altersquotient, z. B. 505
Amtliche Statistik 20, 487, 554, 676, 973, 977, 1171, 1255, 1283, 1342, 1343, 1433
    Daten 161, 281, 539, 584, 591, 967, 1079, 1202, 1255

und internationale Kooperation 487
Weiterverarbeitung der Daten 1283
Analysis:
    of binary data, e.g. 227, 288, 864
    of covariance 21, 93, 100, 210, 251, 272, 285, 286, 412, 573, 585, 674, 732, 751, 780, 857, 911, 955, 1073, 1093, 1095, 1121, 1122, 1128, 1142, 1209, 1219, 1233, 1238, 1253, 1284, 1390, 1398, 1408
    see also ANCOVA
    see also fixed effects
    analysis of the association between cigarette smoking and the age of natural menopause 751
    and standardization as instances of prediction 751
    and the z-score method of adjustment 681
    assumptions: see fixed effects …
    Bayesian, from comparing changes 1219
    consequences of failure to meet assumptions underlying the fixed effect 412
    multiple comparison procedures in the 152
    multivariate 585, 1209, 1284
    on the power of robust tests in 100
    rank and robust 100, 412, 585, 764
    versus stratification 674
    of cross-classifications: see there
    of means 945, 946, 1001, 1086
    of mean scores 374
    of nonequivalent groups 585
    of observational studies 21, 205
    of panel data: see panel analysis
    of residuals 219, 235, 249, 300, 547, 651, 761, 765, 915, 916, 955, 956, 1316, 1340, 1383, 1406, 1437

Analysis:
  of two-way tables by medians, e.g.
    547, 871, 916, 1320, 1339
  of unbalanced data 789, 1142
  of variance, e.g. 83, 93, 285, 288,
    412, 517, 658, 716, 732, 895,
    911, 955, 1001, 1093, 1238,
    1259, 1390, 1408
    see also Varianzanalyse
    and problems under time series
      models 716
    assumptions: see fixed effects
      analysis of variance
    balanced and unbalanced
      designs, e.g. 1259
    components of variance 732,
      780, 1156, 1390
    contrasts see there
    estimability, balanced and un-
      balanced designs, and
      effect definition 1259
    fixed effects 362, 412
    grouping means, type I error
      rates for divisive clustering
      methods for 170
    influential observations in the
      1043
    interactions 131, 232, 417, 525,
      674, 716, 1122
    mixed models 31, 264, 285, 896
    multiple comparisons: see there
    multivariate 12, 285, 482, 587,
      634, 661, 688, 716, 780, 798,
      835, 840, 855, 900, 972,
      1209, 1284, 1323
    nonorthogonal 239, 780, 900
    of nominal data, e.g. 993
    of time-structured data, uni- and
      multivariate 954
    one-way, with time-series data,
      successive observations on
      individual subjects classi-
      fied into different groups
      1432
    parametric alternatives to the
      203
    resolving current controversies
      in 1259

    robust, e.g. 755
    tables for testing ordered
      alternatives in an …
      without replications 734
    two-way classification with un-
      equal numbers of observa-
      tions per cell, e.g. 681, 780,
      789, 1238
    unbalanced, e.g. 681, 732, 780,
      789, 1001, 1043, 1238, 1259
    unknown and unequal variances
      102, 277, 305, 698, 755,
      1288, 1397
    variance components analysis,
      e.g. 732, 780, 1156, 1390
    violations of assumptions 203,
      896, 1288
    with messy data 896
    with unequal variances 102, 277,
      305, 698, 755, 1288, 1397
ANCOVA:
  see also analysis of covariance
  alternatives to 100, 412, 459, 762,
    1284
  estimating the intraclass correla-
    tion coefficient 1253
  on the power of robust tests in
    100
  versus blocking 857
  versus matching and stratification
    21
Andrews plots of multivariate data
  342, 351, 634, 840
Angewandte Statistik: Beispiele, z.B.
  236, 568
Angular variables, association
  between 364
Animal:
  abundance, estimation of … and
    related parameters 966, 1000,
    1064, 1208, 1210, 1391
  carcinogenesis trials, animal
    carcinogenicity experiments
    173, 225
  populations: wildlife sampling
    966, 1000, 1064, 1208, 1210,
    1391

ANOVA:
  see also analysis of variance and
    Varianzanalyse
  and MANOVA:
    comparison of some computer
      programs 716
    graphical methods for internal
      comparisons 716
    inference and the structural
      model for 716
    models for categorical data 716
    robustness of test procedures 716
  components of variance 1156, 1390
  crossed classifications, e.g. 1390
  estimability, balanced and
    unbalanced designs and
    effect definition, e.g. 1259
  of nominal data, e.g. 993
  programs, the specification of
    experimental designs to 517
  regressions in treatment effect
    designs, e.g. 1390
  unbalanced 780, 789, 1043, 1259
Anpassung loglinearer Modelle, z.B.
    115
Anpassungstests 41, 643, 658, 793,
    821, 836, 1049, 1072, 1152, 1153,
    1290
Ansteckende Verteilungen, z.B. 298
Anteilschätzungen in Zufallsstich-
    proben: siehe Stichprobenaus-
    wahl
Antinomien und Paradoxien 882, 884
Antithetic variates 760
Anwender-Programme und Pro-
    grammsysteme 83, 93, 114, 285,
    376, 639, 786, 891, 897, 918, 926,
    1001, 1121, 1122, 1129, 1284, 1412
APC models, age, period and cohort
    models, specification and
    implementation of 851
APL 24
Applied statistics:
  algorithms, e.g. 453
  examples, e.g. 236, 568
Approximation, the accuracy (order)
    of an, e.g. 103

Arbeitsmarktstatistik 161, 1255
Arbeitstechniken, sehr elementare
    Methoden, z.B. 1444
Arbeitszeitstatistik 1079
Archeological data, interpreting, e.g.
    47, 596
Area, estimation of, e.g. 1333
Areal correspondence and areal
    correlation, coefficients of, e.g.
    1333
Area sampling, e.g. 621, 1061
Areas on maps: measuring their
    shape by the compactness ratio
    and the radial line index, e.g.
    1333
ARIMA:
  and ARMA models and processes
    4, 122, 185, 186, 255, 285, 557,
    720, 825, 826, 859, 865, 985,
    1019, 1020, 1183
  models:
    autoregressive integrated
      moving average models:
      see Box-Jenkins fore-
      casting
    to assess an impact of external
      interventions on time series
      865
Arithmetic mean:
  and other location estimators, e.g.
    547, 658
  James-Stein estimator, e.g. 327
ARMA:
  model, e.g. 185, 186, 230, 720, 859,
    1019
  model for successive observations
    over time on individual sub-
    jects classified into different
    groups: testing the equality of
    the group means by the usual
    analysis-of-variance test 1432
  models:
    autoregressive moving average
      models: see also ARIMA
      and ARMA models
    intervention problems and
      outlier detection 720

ARMA:
    models:
        robustness in time series and
            estimating 720
Arrangements and selections, general
    counting methods for, e.g. 1315
Arrhenius equation, e.g. 160
Arrival processes, e.g. 760
Artificial intelligence:
    and expert systems: see expert
        systems
    in analytical chemistry 377
Arzneimittel: Entwicklung und
    Prüfung 160, 1353
ASA-II-Verfahren 1265
Asbestos:
    and other hazardous chemicals,
        e.g. 198
    mining and mortality 791
    workers and cigarette smoking 1091
Ascertainment, weighted distribu-
    tions arising out of methods of 35
Asking questions 125, 167, 564, 609,
    644, 1200, 1276
ASN Function and efficiency 483
Association:
    and causation 21, 101, 137, 169,
        220, 425, 522, 603, 662, 822,
        926, 1137, 1152, 1153, 1182,
        1230, 1277, 1361, 1404
    and regression procedures, non-
        parametric graphical 363
    and mean ridit 611
    between angular variables 364
    in multiway cross-classifications
        having ordered categories,
        e.g. 10, 204, 426–428
    measures of 9, 10, 11, 81, 103, 210,
        263, 349, 429, 611, 724, 739,
        766, 792, 793, 798, 810, 836,
        1110, 1154, 1334
        for contingency tables, e.g. 9,
            103, 204, 352, 353, 355, 367,
            425–429, 766, 792, 836,
            1110, 1334
Associations:
    measurement of, in epidemiology
        21, 338, 682, 1429

weak, in epidemiology 1429
Association within pairs of species,
    e.g. 1065
Assoziationsmaße 9, 10, 81, 103, 210,
    263, 349, 429, 611, 739, 766, 793,
    798, 836, 1110, 1154, 1334
Assumptions of statistical methods
    (e.g. 126, 412, 634, 1238, 1284,
    1323) see robustness in statistics
    and nonparametric statistics
Asymptotic relative efficiency, e.g.
    836, 1072, 1090
Asymptotic theory of point estima-
    tion, e.g. 774
Attack, fake or real: signal detection
    theory applied in the context of
    a two-sided game with reliable
    and deceptive information used 128
Attributable risk 137, 338, 367, 1363,
    1364, 1429
    estimating the population . . . for
        multiple risk factors using case-
        control data 151
Attribute:
    life test data, estimation methods
        based on 847
    sampling plans 483, 781, 984, 1181
Auditors, practical statistical
    sampling for 1395
Aufspaltung von Mischverteilungen
    224, 343, 869
Aufteilung und Schichtung, Op-
    timierung in der Stichpro-
    bentheorie durch 1191
Auftragsabwicklung 1052
Ausfallhäufigkeiten, z.B. 99, 420, 552
Ausgleichsfunktion, Anpassung an
    beobachtete Werte 249, 891,
    1041, 1376
Ausländer in der Bundesrepublik
    Deutschland: Daten 161, 584,
    591, 1079, 1255
Ausreißer 49, 71, 75, 253, 350, 511,
    515, 607, 733, 1135, 1264, 1424
Außenhandel: Daten 1255
Auswahl:
    Geburtstagsstichproben 1167

Namensstichproben 1166
beim Kaufverhalten, z. B. 347
Auswahlverfahren bei Erhebungen
98, 727, 1061, 1169
Auswahl:
von Grundgesamtheiten 74, 307,
406, 407, 465, 466, 1093
von Personen 1166, 1167
Autocorrelation:
against randomness, test of 688
functions, e. g. 4, 122, 185, 186
partial, e. g. 4, 122
Autocorrelations, e. g. 4, 83, 121, 122,
650, 955, 956, 1420
Autocorrelation, the price of
ignorance of the ... structure
of the errors of a regression
model 650
Automated statistical consultation
and data analysis 478
Automatic:
classification: see cluster analysis
interaction detection, e. g. 513
Autoregression, e. g. 185, 186, 824
Autoregressive:
integrated moving average
(ARIMA) models: see
Box-Jenkins forecasting
models (AR) and moving average
(MA) models: see ARIMA
and ARMA models and pro-
cesses
moving average models, Bayesian
analysis of 143
moving average processes, e. g. 4,
122, 859, 958, 1019
processes, e. g. 4, 122, 185, 186
Bayesian analysis of 143
spectral estimation 122, 718, 1080
Autosomal linkage, e. g. 331
Availability:
and reliability, Bayesian 847, 1309
analysis in reliability engineering
524, 847, 949, 952, 999, 1130,
1309
bias in social perception and interaction 640

of maintained systems, e. g. 847
Averaging, reciprocal, e. g. 965
Axial designs, e. g. 224

**Backgammon and other dice diver-
sions** 1011
Banzhaf index: application to
weighted voting in Nassau
County 128
Bar and column charts, e. g. 1188,
1189
Bartlett-Lewis process and Neyman-
Scott process, e. g. 233
Bartlett's test:
critical values for unequal sample
sizes 245
on the determination of critical
values for 316
Bartlett's three-equal-groups resistant
line, e. g. 547
Bartlett-Test:
auf Varianzhomogenität: Bestimmung kritischer Schranken
316
auf Varianzhomogenität: kritische
Schranken für ungleiches n
245
Barycentric coordinates 56, 915
BASIC-Programme, z. B. 891, 918,
1339
Batch arrivals, e. g. 760
Bathtub curve, e. g. 99, 420, 544
see also hazard rate function and
survival data analysis
Batting averages, estimating, e. g. 175,
327
Battles, processes that influence the
dynamics and outcomes of
battles 581
Bayes data analysis, empirical, e. g.
175, 327
Bayesian:
analysis:
of covariance 1219
of linear models 143

7

Bayesian:
  analysis:
    of the linear calibration problem
      589
    of variance, e.g. 143, 663, 684
  approach:
    to multivariate structure 47
    to nonparametric statistics 1225
    to the change-point problem
      1438
  confidence intervals, parametric
    empirical 123
  discrimination 717
  inference:
    in reliability 847
    on the odds and risk vatios 569
  methods:
    of inference, the robustness of:
      estimating the total in a
      finite propulation using
      transformations to
      normality 123
    robustness of 498
  multivariate analysis, e.g. 143, 1078
  procedures, nonparametric 498
  reliability and availability 847,
    1309
  sampling plans 483, 781
  statistical analyses in legal pro-
    ceedings, the presentation of,
    e.g. 354
  statistics 47, 48, 123, 143, 354, 383,
    498, 569, 587, 589, 663, 684,
    769, 801, 847, 915, 1056, 1078,
    1121, 1146, 1147, 1219, 1225,
    1252, 1298, 1309, 1438
Bayes:
  risk and Bayes decisions, e.g. 66
  rules:
    and partitions 132
    frequency of 123
Bayessche Hypothesenprüfung und
  Signifikanztest z.B. 48, 1147,
  1298
Bayes-Statistik:
  als Alternative zur klassischen Sta-
    tistik, z.B. 48, 684, 1146

Ein-, Zwei- und k-Stichproben-
  Fall 684
Bayes's theorem, nomogram for 345
Bayes-Verfahren 47, 48, 117, 123, 143,
  354, 383, 498, 569, 587, 589, 663,
  684, 769, 801, 847, 915, 1056,
  1078, 1121, 1146, 1147, 1219,
  1225, 1252, 1298, 1309, 1438
Bedienungssysteme 193, 339, 347, 695
Befragung 1953 und Wiederholungs-
  befragung 1979: Arbeit und
  Leistung, Lebensqualität, Wand-
  lung der Frauen, Abwendung
  von der Kirche sowie Reisen als
  Leidenschaft 967
Befragungen 117, 125, 167, 564, 609,
  644, 1200, 1276
Behavior:
  and development, longitudinal re-
    search in the study of 954
  change: single-case designs 413,
    527, 571, 1047
Behrens-Fisher Problem 368, 840,
  1183, 1288
Beispiele zur angewandten Statistik,
  z.B. 236, 568
Beratung, statistische 888, 1149, 1257
Berliner Saisonbereinigungsverfah-
  ren 1265
Bernoulli numbers, e.g. 1028
Beschreibende Statistik z.B. 43, 81,
  349, 1151–1153, 1352, 1374
Best-fitting subsets in multiple regression, e.g. 300, 549, 550, 890
Best candidate selection procedures
  406
Bestimmtheitsmaß 451, 1152
Bestimmungsgrenzen in der analy-
  tischen Chemie 244, 320, 577
Best subset selection in multiple re-
  gression, e.g. 300, 549, 550, 890
Beta:
  -binomial:
    distribution and empirical Bayes
      estimation, e.g. 175
    model and its extensions 1396
  distribution 14, 304, 343, 504, 541,
    631, 661, 684, 760, 1186

evaluating and inverting the incomplete beta function 661

highest posterior density intervals 684

function: see functions

tail probabilities: a normal approximation 14

-Verteilung 14, 304, 504, 541, 631, 684

Bevölkerung:

der Bundesrepublik Deutschland nach Geschlecht, Alter, Staatsangehörigkeit und Familienstand 161, 1255

Regional-, Vital- und Sozialstruktur sowie die räumliche Bevölkerungsbewegung 340

Bevölkerungsbewegungen 37, 161, 340, 455, 591, 1148, 1255, 1281

Bevölkerungsgeographie 37, 340

Bevölkerungsprognosen 340

Bevölkerungsprozeß und -prognose 340

Bevölkerungsstatistik 20, 37, 161, 340, 382, 415, 505, 519, 521, 539, 553, 591, 601, 626, 667, 676, 750, 803, 818, 851, 966, 967, 1078, 1079, 1102, 1148, 1171, 1197, 1202, 1255, 1281, 1404, 1409, 1415, 1428

siehe auch Bundesrepublik Deutschland: Daten

Bibliographie 340

ältere Menschen 161, 340, 539, 818, 851, 1148, 1255, 1415

Randthemen 161, 967, 1148, 1255

und Gesundheitsstatistik, die Entropie als Maß in der 692

Bevölkerungsstruktur 161, 340, 505, 1079, 1255

und Bevölkerungsprozeß, z. B. 340

und Bevölkerungswachstum, z. B. 505, 521

Bevölkerungswissenschaft 505, 1148

Bevölkerung und Gebiet: Daten 584, 1148, 1255

Bewertung neuer Methoden in der analytischen Chemie 172, 289, 1387

Bias:

and reliability in the comparison of several measuring methods, restricted maximum likelihood estimation of 172

competition bias, stopping-rule bias, and omission bias in the selection of subsets of regression variables 890

confounding and weak associations in epidemiology 450, 1429

nonresponse: see there

reduction 21, 172, 205, 367, 506, 568, 640, 682, 726, 1074, 1131, 1182

by analysis of covariance 21, 205, 857

by blocking 21, 205, 857

by matching and stratification 21, 205, 857

-Verminderung:

durch Kovarianzanalyse 21, 205, 857

durch Blockbildung 21, 205, 857

Bible, faith versus rationality: game theoretic interpretations of sacrifice in the Old Testament 128

Bibliographical hints concerning statistics and statistical methods 701, 1153

Bildung, Kultur und Wissenschaft: Daten 161, 584, 1079, 1261

Bilinear models 417

Binary data analysis, e. g. 227, 288, 864

Binomial:

and hypergeometric sampling plans 984

-beta distribution, e. g. 542

data plot 546

distribution:

and empirical Bayes estimation, e. g. 175

and related distributions: see discrete distributions

Binomial:
  distribution:
    and selection problems 406
    and statistical inference, e. g. 1072
    bounded: estimation of its
        parameter 748
    confidence limits for propor-
        tions, e. g. 105, 984, 1426
    effects of redundancy, partial
        output states, unavailability
        and one unit in reserve,
        non-identical capacities
        and non-identical
        availabilities, e. g. 99
    statistical inference, e. g. 1072
    testing the goodness of fit of the
        1290
  estimation: group testing 812, 1051
  parameters, improving the normal
        approximation when
        constructing one-sided
        confidence intervals for 486
  -Poisson mixture distributions 343
  populations, inference on trends in
        several 58
  proportions:
    comparing two independent:
        see fourfold table analysis
    confidence intervals for the ratio
        of two 697
Binomialtest 214, 766, 836, 1072
Binomial variables, tolerance interval
    for 479
Binomialverteilung 504, 631, 632, 658
  Anpassungstest, z. B. 1290
  verallgemeinerte 1290
  Vertrauensbereich, z. B. 105, 984,
    1426
Bioassay 160, 251, 359, 360, 531, 582,
    1095, 1332
Bioavailability studies 160, 1353
Biological
  populations, sampling, e. g. 223
  studies concerning environmental
    topics 447
Biometrical genetics, e. g. 331, 852
Biopharmaceutical statistics, e. g. 88,
    160, 1353

Biosignalanalyse, Computerpro-
    gramme zur 371
Biplot 47, 342, 435, 513, 755
Birth and death processes, e. g. 53,
    193, 454, 1136
Birthday problem 687, 1400
Births and deaths 667
Biseriale und punktbiseriale Korrela-
    tion 210, 739, 798, 1122
Bisquared function, e. g. 871
Bivariate:
  continuous distributions, e. g. 643
  data, methods for investigating 47
  generating functions, e. g. 643
  linear correlation 798
  normal:
    association models for contin-
        gency tables with ordered
        categories 427
    distribution 427, 446, 643, 798,
        875, 1008, 1029
    probabilities, a simple approxi-
        mation for 875
    screening model 1008
    tolerance regions 485
  Normalverteilung 643, 798, 875,
    1008, 1029
  semi-lognormal distribution, esti-
    mation in a 1278
  survivor functions, e. g. 234
Biweight estimators, e. g. 547, 871
Blending:
  chemical pesticides for control of
    mites 224
  experiment, seven-component
    octane 224
Blinding (in clinical trials) methods
    and measurements of, e. g. 1214
Blockbildung gegenüber Kovarianz-
    analyse 857
Block design: model I and mixed
    models 1036
Block designs 251, 552, 674, 1036,
    1107, 1257
  and distributionfree tests on
    aligned observations, e. g. 836
  and tests on aligned observations
    836

complete, rank tests for 509
incomplete 251, 1257
randomized 121, 674, 955
    treatment versus control, exact
      tables of the distributions
      of rank statistics 980
Blocking:
  and randomization 121, 205, 270,
    1151–1153, 1238
  and restricted randomization 40
  orthogonal, e.g. 120
  versus ANCOVA 857, 1284
Blocks, randomized, based on
    alignment of the original
    observations: distribution-free
    tests, e.g. 836
Block-Versuche 251, 552, 674,
  1107
Blom's plotting position, e.g. 808
BMDP-Programme 83, 114, 155, 285,
  404, 736, 926, 1017, 1196, 1264,
  1284, 1437
Bonferroni:
  chi-square statistics, e.g. 1153
  -Holm-Prozeduren 59, 565, 1141,
    1180
Bonferroni's and Hunter's inequality
  59
Bonferroni-Statistiken und -Tests 38,
  59, 391, 565, 712, 794, 856, 914,
  955, 1089, 1180
Bootstrap:
  methods 269, 325, 326
  quantile, e.g. 1025
Bounded binomial distribution and
    bounded multinomial distribu-
    tion 747, 748
Bowker-Test 709, 836, 1153
  Verallgemeinerungen 1449
Box:
  and-whiskers plot: see box plot
    variations
  -Cox transformations, e.g. 547,
    716, 916, 1052, 1320, 1390
  -Jenkins forecasting 4, 122, 255,
    285, 557, 720, 825, 826, 985,
    1019, 1020, 1184

-Muller method for generating
    pseudo-random normal
    deviates 421, 1035
plot variations 23, 502, 547, 866,
  871, 1001, 1024, 1319, 1320,
  1339, 1359
Boyce-Codd-Normalform 576, 1330,
  1348
Branch and bound method 1308
Branching processes, e.g. 454, 874,
  1136, 1232
Breakdown:
  bound breakdown point 488, 547,
    580, 1097
  point, breakdown bound 488, 547,
    580, 1097
Breast cancer survical, e.g. 437
Breslow/Gehan and logrank
    methods for survival analysis
    134, 137, 138, 144, 770
Brownian motion, e.g. 193, 233, 347,
  454, 874, 1136
Brown-Mood median regression line,
  e.g. 363, 547
Bruchpunkt: siehe breakdown bound
  oder breakdown point
Bruttosozialprodukt 20
Bulk sampling, e.g. 1181
Bundesamt, statistisches: Aufgaben
  487, 554, 973
Bundesrepublik Deutschland: Daten
  161, 162, 163, 281, 415, 539, 553,
  584, 591, 967, 1079, 1202, 1255
Bureaucracies, statistical practice in
  846
Burr:
  distribution 541, 1285
  type XII and related distributions
    1285
Butler statistic 290

**Cabinet coalition formation, a game-
    theoretic analysis** 128
Cadmium and zinc: distribution in
  blood 1278

11

Calibration 148, 300, 318, 319, 377, 395, 577, 589, 610, 685, 733, 815, 832, 990, 1176, 1280, 1295, 1324
  analyzing residuals in . . . problems 990
  inverse median estimation problems, new simulation results for 1324
  choosing between classical and inverse estimators 815
  comparative . . . and congeneric measurements 610, 1295
  congeneric measurements 610, 1295
  curve: see calibration
  curves and inverse regression 300, 832, 990
  decision and detection limits 577, 1280
  inverse median estimation 1324
  linear 318, 577, 815, 1280, 1295
    Bayesian analysis 589
    compound estimation 815
  multivariate 148
  nonconstant variance regression technique 395
  nonlinear 319
  nonparametric 685
  of probabilities 640
  problems, analyzing residuals in, e.g. 990
  statistical theory of 1176
Caliper matching, e.g. 682
Cancer:
  see also tumor and carcinogenesis
  avoidable risks 198, 225, 292, 338, 791, 888, 917, 1021
  (possible) causes of 198, 225, 292, 338, 791, 917, 1021
  chemotherapy: regression analysis of survival data 173
  data, application of nonparametric statistics to 719
  prognosing 612
  research:
    optimal dose estimation in cancer chemotherapy 173
    statistical methods 225, 612, 888, 1021
    risks of, quantitative estimates of avoidable, e.g. 198, 292
    screening for the early detection of 225
    survival studies: examples 225, 437, 649, 1060
Canonical:
  correlation analysis 8, 22, 93, 114, 285, 411, 482, 501, 658, 661, 686, 731, 767, 798, 800, 835, 840, 855, 922, 1078, 1284, 1303, 1316, 1350, 1381
  forms, e.g. 1207
  variates: see canonical correlation analysis
Capacity outage (available) probability tables and their use in reliability evaluation, e.g. 99
Capture-recapture models 103, 966, 1000, 1064, 1065, 1208, 1210, 1391
Carcinogenesis 338, 713, 917, 1021
Carcinogenicity:
  bioassay, design and analysis of the long-term 198
  studies, examples based on animal 173, 198, 225
Carcinogenic risk assessment 84, 137, 225, 361, 418, 713, 917, 1021, 1182, 1231
  see also attributable risk, odds ratio and relative risk
CARE III, Computer-aided reliability estimation III, e.g. 46
Carry-over:
  effect in cross-over designs, see cross-over designs
  effects, e.g. 1408
Case-control:
  data: conditional and unconditional likelihoods 813
  study 6, 86, 134, 135, 137, 139, 142, 151, 225, 258, 338, 367, 561, 791, 813, 1182, 1230, 1429
    see also odds ratio
    control group selection by sampling of telephone households 497

12

cost of 142
general relative risk functions
for 139
overadjustment in 258
relative risk estimators 561
Catanova and the analysis of variance of nominal data 993
Catcher matrix 916, 1316
Categorical data:
analysis 103, 334–336, 352, 353, 355, 374, 425, 490, 519, 520, 658, 663, 766, 798, 814, 822, 833, 965, 1056, 1170, 1425
assessment of dispersion in 1170
causal analysis of, pitfalls in 353, 425, 822
cross-classified: see contingency table analysis or log-linear models
dispersion in 1170
in longitudinal studies of behavioral development, the analysis of 954
Light-Margolin measure of association in contingency tables 83
model building 103, 352, 353, 374, 425, 519, 798, 1110, 1122, 1334
problems using information theoretic approach 719
ranked, assigning quantitative values to: see ridit analysis
recent advances in theory and methods for the analysis of, e.g. 353, 612
sequential, tests of dominance in 1365
Categorical variables:
ordered, regression and 18, 862
quantification of, e.g. 612
Categoriced data: see categorical data analysis
Category matching, e.g. 682
Cauchy-Verteilung 504, 631, 1127
Causal:
analysis:
in sociology 101, 425, 845

of categorical data, pitfalls in 352, 822
diagrams 522
inference: nonexperimental 220, 845, 1404
modeling and survey research, e.g. 1137
models, e.g. 21, 101, 137, 169, 220, 425, 522, 603, 662, 682, 822, 926, 1119, 1137, 1182, 1277, 1361, 1404
and true experiments, e.g. 662
evaluating: an application of maximum-likelihood analysis of structural equations, e.g. 848
in longitudinal research 954
with multiple unmeasured variables, e.g. 662
with unmeasured variables, e.g. 662
systems, and causal models with time series data, e.g. 169
Causality versus correlation, e.g. 662, 1230
Causation, causal models, and causal analysis, e.g. 21, 101, 137, 169, 220, 425, 522, 603, 662, 682, 822, 926, 1119, 1137, 1182, 1277, 1361, 1404
Causes:
and effects of nonresponse in surveys 205, 594, 805, 820, 928, 1153
in epidemiology 682, 1139
of cancer 84, 292, 713, 888, 1021
of death 166, 818, 1148, 1255
among the elderly as health indicators 818
Censored:
data, e.g. 719, 770, 952, 1010
comparing two sets of 21, 694, 770, 834, 893, 1291
failure data hazard plotting 950, 952
survival data: statistical methods, e.g. 134, 137, 560, 642, 649, 651, 761, 762, 770, 834, 893, 950, 1010, 1177

Censoring, types of 134, 234, 761, 762, 770, 893
Census data 667, 1404, 1433
Censuses of animal populations 966, 1000, 1064, 1065, 1208, 1210, 1391
Census:
plus samples 676
X-11-Verfahren 1265
Central:
limit theorem, e. g. 643, 768, 1028, 1072
nonapplicability to finite populations 1057
statistical office (U. K.) 913
tendency of a pattern of points: mean centre and standard distance 1333
CHAID, Chi-squared automatic interaction detection, e. g. 513
Chain sampling, e. g. 1181
Chance and statistics, e. g. 675
Chancen-Verhältnis, odds ratio 6, 10, 21, 86, 137, 352, 367, 396, 397, 561, 569, 574, 682, 888, 1110, 1182, 1292, 1299, 1447
Change:
analyzing data from premeasure/postmeasure designs 21
functions on the intra- and inter-individual levels, developmental 954
measurement of 103, 422, 1320, 1321
of behavior: single-case designs 413, 527, 571, 1047
-of-value curve 488, 547, 580, 1097
-over designs, e. g. 160
point
detection: see change point problem
problem 59, 513, 1175, 1212, 1438
problem and two-phase regression: an annotated bibliography 1212

relative, the measurement of 1310
social, the cohort as a concept in the study of 851
Characteristic:
functions, e. g. 658, 1028
roots and vectors, e. g. 54, 443
Charts:
and graphs: see also graphical displays and graphical methods
control 924, 1329, 1389
framed-rectangular, e. g. 200
graphs and other types of data display, e. g. 889
two- and three-dimensional, e. g. 1188, 1189
Chebyshev:
inequalities 1028
inequality with estimated mean and variance 1164
Checking the model: see goodness-of-fit and residual analysis
Chemical composition experiments 270
Chemistry:
and statistics 377
detection limits in analytical 244, 320, 577
factor analysis in 827
Chemometrics 377, 827
"Chernoff:
faces" 342, 351, 370, 1359, 1366
-type faces 342, 351, 370, 1359, 1366
clustering with 1305
Chiffren: siehe Kryptographie
Childlessness of marriages in the FRG: a GSK-model application 519
Chiquadrat:
Approximation kritischer Schranken 1442
Kombination unabhängiger Chiquadrat-Tests z. B. 704
Tests 103, 263, 658, 769, 793, 836
-Werte, standardisierte, z. B. 1384
Chi-square:
or chi-squared distribution: algo-

rithm to obtain critical values
191
distribution:
  approximations 158, 545, 1442
  integrating the bivariate: exact p
    values for F and t tests 744
  noncentral 514, 541, 1028, 1032
  percentage points 158, 545, 661,
    1442
  probabilities 661
Chi-squared:
  automatic interaction detection,
    CHAID, e.g. 513
  distribution, noncentral 514,
    541, 1032
  percentage points: approximations
    545
  tests, combining independent 704
  two-by-two table test with $(n$-1) in-
    stead of n in the numerator
    1111, 1151–1153, 1335
Chi-square:
  misuse of 263
  partitioning of 103, 263, 352, 793,
    836, 1100, 1110, 1153, 1334
  statistics, Bonferroni chi-square
    statistics, e.g. 1153
Chisquare
  test:
    fourfold: see fourfold table
      analysis
    fourfold, replacing n by $(n$-1) in
      the numerator 1111,
      1151–1153, 1335
  testing:
    repeated 957
    the equality of two independent
      chi-square values 263
  tests e.g. 769, 836, 1153, 1238, 1334,
    1335
    for 2×2 contingency tables: see
      fourfold table analysis
Chi-Verteilung, logarithmische inverse 684
Cholesky decomposition 661, 1205
Choropleth map e.g. 351, 1188, 1333
  error index for a, e.g. 1333

Choice of indicators in data analysis
915
Chronic disease:
  screening in 910
  risk: logistic multiple regression
    1421
Chrysotile exposure 791, 1091
Circular:
  data:
    a correlation coefficient for 364
    correlation 364
  distribution and distributions on
    the sphere 57, 364, 839, 840,
    1029, 1373
Circularity:
  and/or sphericity, e.g. 44, 975
  assumption associated with block
    designs 674
Circular statistics 57, 364, 839, 840,
  1029, 1373
  in biology 57
CIS, Current Index to Statistics 403,
  478
Claim:
  number process, e.g. 64
  provisions, estimation of
    outstanding . . . in insurance,
    e.g. 575
Classification:
  and clustering procedures, e.g.
    435, 716, 1129
  and discrimination models 537,
    658, 1078
  and grouping techniques, e.g. 634
  and identification: see cluster
    analysis
  and regression trees 132
  functions, evaluating, e.g. 634
  nonparametric, e.g. 717
  of multivariate data: see cluster
    analysis
Classifications, assessment of, e.g.
  435
Classification with several
  populations, e.g. 634
Classifier, assessing a 490
Client-consultant interaction, e.g.
  1149

Clinical:
  chemistry, reference values in 441
  data: ill-defined or unreliable
    variables 160
  studies 437, 649, 888, 1348, 1384
  trials 96, 112, 160, 694, 780, 788,
    797, 878, 888, 1048, 1058,
    1059, 1214, 1326, 1348, 1354,
    1393
    adaptive allocation procedures
      for 28, 55
    adverse reactions, e.g. 788
    allocation of treatments on the
      basis of outcome 28, 55
    checklists for 96, 160, 1214
    correspondence analysis in 612
    multi-center 112
    of analgesic drugs, e.g. 160
    of psychoactive drugs, e.g. 160
    of treatments with dichotomous
      outcome: paired versus
      two-sample design, power
      considerations 1354
    optimal allocation of treatments
      in sequential 28, 55
    optimality in 28
    sample size estimation in the
      design of, e.g. 296
    selection of trial patients, e.g.
      160
    sequential:
      methods for 28, 55, 1393
      on the allocation of treat-
        ments in 28, 55, 1393
    the incomplete inpact on future
      medical practice of results
      28, 55
    the search for optimality in 28
    use of confidence intervals:
      effect of beta-blockers on
      post-infarction mortality
      788
Closed population, estimating the
  size of a 103, 966, 1000, 1064,
  1065, 1208, 1210, 1391
Closeness of fit: see goodness-of-fit
  and residual analysis

CLUSTAR/CLUSTID package
  1129
Clusteranalyse und Diskriminanzanalyse: Beispiele, Methoden
  und Programmpakete 262
Cluster analysis 8, 83, 108, 114, 285,
  342, 346, 435, 442, 482, 490, 501,
  513, 608, 612, 715, 717, 810, 840,
  844, 855, 869, 907, 1078, 1085,
  1129, 1195, 1209, 1245, 1258,
  1305, 1323, 1412
  algorithms for determining
    partitions and overlapping
    clusters, e.g. 608, 717
  and factor analysis: FORTRAN
    programs, e.g. 608
  automatic interaction detection:
    see there
  distances and proximities in auto-
    matic classification, e.g. 608
  hierarchical classifications, e.g.
    608
  optimization methods of 844
  resemblance coefficients for
    quantitative and for qualita-
    tive attributes and for mix-
    tures of quantitative and
    qualitative attributes, e.g.
    1129
  software, e.g. 717, 1129
Clustering:
  approaches by Day and Scott and
    Symons, a comparative study
    of the 612
  methods:
    and pattern recognition 342, 612
    hierarchic and non-hierarchic,
      e.g. 513
  problems and recent advances, e.g.
    612, 634, 717
  with Chernoff-type faces 1305
Cluster:
  process, e.g. 233
  sampling 205, 621, 645, 729, 1061,
    1137, 1271, 1274, 1395, 1431,
    1433
  validity of fuzzy clustering
    algorithms 612

Cochran's test on the homogeneity of variances, a Monte Carlo computer program for the critical values of 1144

CODASYL, Conference on Data Systems Languages, e.g. 576, 1330

Coded tables, e.g. 1339

Coding of data and code construction, e.g. 1243

Coefficient:
of correlation: see correlation coefficient
of variation:
adequacy of the chisquared approximation 1370
models for data with constant, e.g. 864
plots 566

Cohort:
analysis 35, 47, 138, 161, 194, 282, 338, 367, 416, 422, 682, 791, 851, 1182, 1230, 1232, 1292, 1404, 1428
application to asbestos mining 791
elementary methods of 136
in epidemiology: see cohort analysis
in social research: beyond the identification problem 851
multiplicative models 138
as a concept in the study of social change 851
change, dynamic modeling of: the case of political partisanship 851
effects in developmental psychology 954
excess mortality of men in Germany 282
-inversion models 851
life table e.g. 194
for Germany, a tentative computation of a 282
period and age models and effects in demography 851

studies, multiple, retrospective and prospective 851
study/analysis (follow-up study) 47, 138, 161, 194, 225, 282, 332, 338, 367, 416, 422, 682, 791, 851, 1182, 1230, 1292, 1404, 1428
study: analysis of stratified incidence or mortality data 1292

Cohorts:
generations, and conformity 851
several: simultaneous analysis of longitudinal data from 851

Collaboration and consulting 350, 477, 478, 888, 1149, 1257, 1441

Collaborative test, e.g. 733

Collapsing:
contingency tables 16, 103, 352, 353, 425, 513, 1110
variables:
in contingency tables: see collapsing contingency tables
in log-linear models, e.g. 1110

Collecting data, e.g. 889

Collinearity 75, 551, 831, 1316, 1340, 1383, 1402
and predictability, e.g. 1402

Color statistical maps 351

Combat:
in fiddler crabs, game theory models of intermale 128
parameters, the impact of warfare models on 581

Combination of independent tests 414, 588, 719

Combinatorial formulas, combinatorials, some useful, e.g. 1028

Combinatorics 719

Combining:
and comparing significance levels from nonindependent hypothesis tests 1270
and contrasting significance levels 1270
comparable test results, e.g. 414, 588, 719, 1132, 1153, 1270, 1362

Combining:
  evidence from fourfold tables 137, 227, 367, 1152, 1153
    see also Mantel-Haenszel summary odds ratio
  independent: chi-squared tests 704
    F tests, Bayes procedures for 650
  variables to represent a kind of supervariable: see e.g. multiple regression analysis, canonical correlation analysis, and principal components analysis
  probabilities, e.g. 414, 1132, 1153, 1270, 1362
  significance levels: procedures necessary for adjusting for nonindependence 1270
Communication:
  and mathematics, e.g. 606
  network analysis methods 900
  processes, stochastic modeling of 900
  research, human, multivariate techniques in 900
Community risk assessment:
  attributable risk 137, 338, 367, 1363, 1364, 1429
Comparative:
  assessments, economic and social statistics for 35
  calibration and congeneric measurements 610, 1295
  experiments, the design of: see experimental design
  studies, statistical methods for, e.g. 21
  trial, tests for the 2×2: see fourfold table analysis
Comparing:
  alternative systems, statistical techniques for, e.g. 760
  (for integrating studies) experimental and control conditions: the probability of failing to detect an effect size 414

pre-event and post-event time series segments to estimate the magnitude and form of the impact 865
sequences, e.g. 374, 433, 434, 439, 440, 597, 766, 775, 793, 836, 1153
several censored survical curves 134, 137, 138, 649, 651, 761, 770, 950
several treatments with a control, distribution-free tests for 369
survival distributions by nonparametric methods: two- and k-sample cases with programs 770
two groups on p variables 800, 1089
two sets of arbitrarily censored data 770, 834, 893
Comparison:
  of ANOVA programs 517
  of five allocation rules for allocating treatments in sequential medical trials 55
  of groups of observations by ridit analysis, e.g. 367, 611
  of period and cohort life tables 194, 282, 1428
  of rates or proportions 296, 916
  of several measuring methods, restricted maximum likelihood estimation of bias and reliability in the 172
  of the iterative percentile and the maximum likelihood estimation for a normal-normal mixture 149
  of three regression lines over a finite interval 530
  of two independent samples or two matched samples of response curves, nonparametric 708
  of two or more groups of patients adjusted for normal mortality expectation 145
  of within-block rankings of not necessarily equal blocks, a distribution-free test 94

plots or schematic plots, e.g. 871, 1320
Comparisons, paired: 719
bibliography on the method of 254
Comparison with a control 406, 466
Comparisons with a control or standard, multiple, e.g. 306, 309
Competing risks, e.g. 234, 642
Competition bias in best subset selection 890
Competitive interaction between two species, e.g. 1065
Completely randomized design, e.g. 674, 955
Complex demodulation 718
Compliance sampling, e.g. 1181
Component:
  analysis in chemistry: application of factor analysis 827
  state resident times: systems with exponential distributions, and systems with non-exponential distributions, e.g. 99
Composites, selection of 1323
Compounding probabilities from in-dependent significance tests 414, 1132, 1362
Compound:
  Poisson process, e.g. 64
  probabilities: bounds for correlated 1377
Computational:
  geometry and statistical applica-tions 97
  statistics:
    definition, history and future 1346
    see statistical computing
Computation:
  and mathematics, e.g. 606
  numerical, difficulties in 520
Computer:
  -aided:
    experimental design 1257
    reliability estimation III, CARE III, e.g. 46
  generation of samples from

exponential, normal, gamma, Poisson, ... distributions, e.g. 1117
  graphics, e.g. 97
  -intensive methods in statistics: see jackknife techniques and bootstrap methods
  networks, assessing the reliability of, e.g. 46
  packages: see/siehe Programm-systeme
  program listings: see e.g. BMDP or/and SPSS or/and SAS
  -programme zur Biosignalanalyse 371
  reliability 46, 97
  science and statistics: interface proceedings, e.g. 97
  software, assessing the reliability of, e.g. 46
  -supported instruction: prospects and pitfalls 97
  -Verfügbarkeit 46, 283
Computing:
  networks and computer software, reliability of 46, 97
  new developments in statistical 404
  statistical: see statistical computing and Programmsysteme
  symbolic, and statistics, e.g. 520
Conception and birth: probability models of 1232
Concordance:
  internal, external ordering relative to the, correlation of a group of rankings with an 1018
  of rankings 433, 434, 790, 1018
  see also agreement, rank correlation, and Friedman test
  of two groups of rankings 790, 836
Concurrence and parallelism: tests for ... of two dependent regression lines 1236
Conditional and unconditional likelihoods for case-control data 813

Condition/event systems, e.g. 1105
Confidence:
  ellipses for pairs of mean components, simultaneous, e.g. 634
  interfractile, and interpercentile intervals in laboratory medicine, e.g. 441
  intervals 643, 658, 1090
    for binomial and Poisson parameters, one-sided, improving the normal approximation 486
    for 17 continuous and 4 discrete distributions 1028
    for the variance of a normal distribution 802
    in legal settings 1241
    nonparametric methods, e.g. 1072
    versus significance tests, e.g. 538
  limits:
    for a proportion 486, 984, 1005, 1426
    for a quantile 1005, 1426
    for defective units in a lot 984
    for unavailability e.g. 99
Confirmatory:
  data analysis, CDA 1318, 1322, 1323, 1347
  factor analysis, e.g. 806, 807
    conceptual replication and, e.g. 588
  statistics: see confirmatory data analysis
Conflict models 637, 1092
Confounder, confounding factors 21, 205, 450, 682, 1182, 1429
Confounding:
  and misclassification 450
  and risk factors 682
  bias and weak associations in epidemiology 1429
  conditions for, of the risk ratio and of the odds ratio 113
  essence and detection 21, 682, 887, 1182
  in epidemiology 21, 113, 682, 1182, 1429
  in experimental design, e.g. 674, 1036
  in premeasure/postmeasure designs 21
  overadjustment and matching in epidemiology 135
  variables
    and matching 21, 205, 367
    in observational studies 21, 205
Congeneric measurements, comparative calibration and linear structural relationship 610, 1295
Conjoint analysis 482
Consistency in estimation and in tests, e.g. 547, 658, 1072
Constraints, homogeneous and non-homogeneous, for multiple regression 1115
Consulting:
  and collaboration in medicine 888
  in the physical, chemical, and engineering sciences: experimental design 1257
  statistical 350, 477, 478, 888, 1149, 1257, 1441
  -teaching interaction 1149
Consumer intent, assessing, e.g. 175
Comsumption and income: consumption function, e.g. 1420
Contagious distributions, e.g. 298
Contingency table analysis 24, 93, 103, 114, 115, 123, 214, 263, 285, 288, 334–336, 352, 353, 355, 374, 425–429, 519, 653, 658, 660, 793, 798, 835, 836, 854, 855, 861, 1012, 1026, 1056, 1100, 1101, 1110, 1334, 1335, 1357, 1384, 1425, 1449
  see also GSK approach and loglinear models
  a comparison of three parametric techniques for 1279
  association and independence 9, 83, 103, 204, 425–429, 611, 724, 766, 836, 947, 1334
  exploratory 659, 1320
  interactions in, e.g. 1311

multivariate 103, 352, 353, 355, 835, 855, 1110, 1334

quasi-independence, conditional independence, e.g. 472, 1110

square tables 103, 470, 793, 836, 837, 861, 1056, 1152, 1153, 1334, 1449

tests of symmetry in three-way tables 1101

the relative disequilibrium as a measure of interaction 659

three-dimensional tables 103, 334, 335, 352, 425–429, 472, 836, 1056, 1101, 1110, 1334

three parametric techniques for 1279

Contingency table:

collapsing or 16, 103, 352, 353, 425, 1110

dual scaling 965

exact simultaneous comparisons with control in an $r \times c$ 1026

goodness-of-fit procedures for the evaluation and selection of log-linear models in multilevel tables 115

incomplete: polytomous variables 1334

mean ridits 611

multiway tables and dual scaling 965

of type 2×2 see fourfold tables and Fisher's exact test

ordered categories 10, 204, 263, 1027, 1357

ordered, exact tests for trend in an 1027

partitioning of chi-square 103, 263, 352, 793, 836, 1100, 1110, 1153, 1334

simultaneous comparisons with a control 1026, 1141

Contingency tables:

see also logit models and loglinear models

adjusted residuals 472, 620

adjustment of marginal tables and

adjustment of marginal totals using unsaturated models 472

Agresti's measures of association for 9

and dual scaling 965

and mean ridits 610

cell values, standardized cell residuals 103

estimating cell probabilities, e.g. 103

exact test for $r \times c$ 877, 1011

Fisher's exact test in $r \cdot c$ tables 877, 1012

graphical models, e.g. 323

graphical nonparametric techniques for preliminary analysis of 659

incomplete, e.g. 103, 336, 472

incomplete, inseparable, separable and quasi-independent, e.g. 1334

large sparse tables 323

multidimensional 103, 323, 334, 335, 352, 353, 355, 426–429, 472, 513, 835, 1056, 1334, 1384

multidimensional, graphical models 323

multidimensional: interaction graph 323

ordered, exact tests for trends in 1027

partitioning chi-square 103, 263, 352, 793, 836, 1100, 1110, 1153, 1334

quasi-independence, e.g. 472, 1110

residuals in, e.g. 103, 472, 1110, 1320

reversing decompositions for arbitrary 2×2 and $r \times c$ 884

sampling zeroes in saturated and unsaturated models, e.g. 1110

square, e.g. 103, 470, 1056

square tables with ordered categories, e.g. 861, 1449

square, tests for marginal homogeneity in 710

Contingency tables:
symmetrical multidimensional 103, 472
symmetry in square 103, 1334
test of symmetry in three-way tables 1101
the robustness of a hierarchical model for multinomials and 123
2 × 2: see fourfold table analysis
with ordered categories, analysis of 10, 204, 263, 426–428, 1027, 1357
Continuity correction, Yates's correction 470, 854, 1003, 1004, 1111, 1335, 1434
Continuous:
distributions 504, 541, 631, 701, 1028
multivariate distributions e.g. 631, 712
random variables, e.g. 454, 643
variates, e.g. 454, 643
Contour plots, e.g. 224
Contraceptive effectiveness, measures of 1428
Contrasts:
an improved table for simultaneous control on g 391
linear, of means, testing without assuming normality and homogeneity of variances 1307
monotone, of normal means, simultaneous confidence intervals for 838
ordered: ensemble-adjusted p values 1133
orthogonal, multiple comparisons for 67, 83, 286, 789, 955, 1408
selected and multiple comparisons, an extension and generalization of Dunnett's procedure 1067
simultaneous control on g contrasts 391
weighted as to their importance and interest 1133

Control:
and confrontation: game theoretic models, e.g. 128
and optimization in analytical chemistry 377
charts 924, 1329, 1389
comparing several treatments with:
distribution free tests 369
tables of optimal allocations of observations 68
for extraneous variables in epidemiologic research, e.g. 682
group construction using multivariate matched sampling methods that incorporate the propensity score 1131
Controlled trials: see clinical trials
Control:
multiple comparisons with a control when variances are unknown and unequal 306, 309
schemes, design of discrete, e.g. 122
Controls, how many? 388
Convergence:
of random variables, e.g. 454
types of, e.g. 1028, 1072
Conversational:
packages in statistical computing, e.g. 7
statistics with IDA 1121
Conviction policy for offenses that may have been committed by accident, an optimal 128
Cook's distance, Mahalanobis distance and Weisberg's test: identifying outliers in regression analysis 75, 219, 956, 990, 1264, 1316, 1340, 1383
Coping with deception, a game theoretic model 128
Coronary heart disease risk 1314, 1360
Correlated variates with incomplete data, testing equality of variances of 95

Correlational results: methods for
    summarizing independent 1344
Correlation:
  analysis 8, 210, 658, 739, 792, 798,
      835, 902, 956, 1152, 1153, 1316
  and causation: see association and
      causation
  autocorrelation, see there
  between some statistics 739
  biserial and pointbiserial 210, 739,
      798, 1122
  coefficient
    (or product-moment correlation
        coefficient)
    bootstrap distribution of the
        269
    confidence interval approxima-
        tions 116
    converting r into a standardized
        mean difference 414
    critical values 981, 987, 988
    effect of violation of normality
        on the t test of the 321
    for circular data 364
    for summarizing the informa-
        tion contained in
        probability plots 808
    meta-analytic methods for
        cumulating correlations
        across studies, e.g. 588
    multiple: tables of upper per-
        centage points 771
    Öksoy-Aroian tables 987, 988
    point, e.g. 1065
    robustness of some transforma-
        tions of the sample 1273
    tetrachoric 210, 276, 284, 739,
        798
    transformations of the sample,
        robustness to nonnormality
        of some 1273
    use of the ... with normal prob-
        ability plots 808
  coefficients,
    an empirical study of related
        124
    a robust test for multiple com-
        parisons of 317

  different kinds of 210, 658, 739,
      798, 836, 1110, 1153
  equality of dependent 77, 124,
      934, 1256, 1418
  guidelines for converting
      various summary statistics
      into, e.g. 414
  ordinal, e.g. 1110
  related, a distribution-free test
      for 1418
  related, an empirical study of
      124
  robust test for multiple compari-
      sons 317
  selection procedures for, e.g.
      406
  testing the difference between
      two 1153, 1297
  three related, one-sided test for
      equal correlation 1418
cosine pi, e.g. 739
equal, one-sided tests for ... in a
    trivariate distribution 1418
illusory correlation and the
    reversal paradox 884
inflated with outlying values and
    deflated because of restricted
    range in sampling: remedies,
    e.g. 1284
influence-enhanced scatter plots
    1296
intraclass: see intraclass correla-
    tion
matrices:
  and covariance matrices, e.g. 22,
      634, 1207
    robust, e.g. 580
  examining, by means of proba-
      bility plots 342
matrix:
  missing data, e.g. 1284
  significance tests for comparing
      elements of a 1256
multiple 8, 210, 658, 681, 771, 798,
    840
of change Y-X with initial status X
    662

Correlation:
    partial, e.g. 93, 658, 681, 1238
    or regression: bias or precision
        1369
    phi 739
    polychoric, e.g. 427
Correlations across studies,
    cumulating, e.g. 588
Correlation:
    selecting the normal population
        with the largest intraclass
        correlation and with the
        largest multiple correlation
        coefficient 406
    squared semipartial 1284
    tetrachoric 210, 276, 284, 427, 739,
        798
    versus causality, e.g. 662
Correspondence analysis 47, 164,
    448, 535, 596, 608, 612, 767, 800,
    965
    and related techniques for large
        matrices 767
    versus principal component
        analysis for highly skewed
        distributed variables 337
Cosine Pi correlation, e.g. 739
Counting methods for arrangements
    and selections, general, e.g.
    1315
Counts or/and counted fractions
    1320
Covariance:
    adjustment and standardization as
        instances of prediction 751
    algebra, e.g. 662
    analysis:
        see analysis of covariance
        and prediction, e.g. 751
    and correlation matrices, robust,
        e.g. 580
    function: longitudinal studies, e.g.
        1368
    matrices, e.g. 54, 1207
        according to systems of structur-
        al equations, analyzing: see
        LISREL

    and their partitioning, e.g. 634
    likelihood ratio tests for 716
    likelihood ratio tests for rela-
        tionships between two 650,
        716
    simultaneous test procedures for
        716
    tables for different tests 712
    testing the equality of 31, 712,
        814
    matrix:
        prepresentation and object-
        predicate symmetry 717
        simplifying the structure of the:
        see canonical correlation
        analysis, principal compo-
        nents, analysis and factor
        analysis
    rank analysis of 585, 764
    structure, choosing a model for
        the, e.g. 1368
    structures:
        analysis of, e.g. 513, 634, 807,
        1368
        see principal components
        analysis and factor analysis
Covariates (chosen independent
    variables because of their
    known association with the de-
    pendent variable), evaluation of,
    e.g. 1284
Covariation assessment 640
Cox's
    F-test for comparing two
        exponential distributions 770
    linear logistic regression method as
        an index of risk for a given
        disease, helping to identify
        risk factors, e.g. 770
    models for survival analysis 228,
        234, 285, 770
C_p-statistic 4, 83, 249, 300, 351, 890,
    1383, 1437
Criminal court and statistics, e.g. 299,
    354, 356, 1241
Crop forecasting, statistical problems
    in 35

Cross-classifications having ordinal variables, a survey of strategies for modeling 10

Cross:
classifications: measures of association 103, 425–429, 1110, 1334
-classifications:
ordinal variables: loglinear and logit models, e.g. 9, 10
regression analysis of, e.g. 1110
with ordered categories, association in 10, 352, 353, 426–428, 861, 1027, 1357
-classified categorical data, analysis of 103, 352, 353, 355, 425–429, 1056, 1110, 1334
-over design, cross-over trial 160, 367, 586, 788, 896, 1058, 1326, 1348
-product ratio:
and general functions 103, 352, 659
odds ratio: see there

Cross
-sectional:
and time series data, pooled 274
study 367, 682, 1230
-section surveys, regression models for grouped populations in 1050
-site and within-site analyses, e.g. 889
-tabulated data, analysis of: see fourfold tables, contingency tables, and multiway tables
-validation:
computational methods for generalized 97
procedures for the loglinear model 115

Crude rates: see standardization
Cryptography 91, 92, 187, 606, 696
Cryptology 187
Cumulants:
and cumulant spectra 718

of sampling distributions, e.g. 658
Cumulative:
distribution:
function, empirical, e.g. 643, 658, 1024, 1042
functions, comparison of empirical and theoretical 643
logits, models for, e.g. 11
Curtailed sampling by artificial attributes 781
Curtailment:
as an early stopping rule 406
tables for the multinomial selection problem 406
Curve fitting 249, 891, 1041, 1376
by least-squares (LINWOOD) and by nonlinear least-squares (NONLINWOOD) programs 249
with smooth functions that are piecewise-linear in the limit 1082
Curves:
common . . . and the transformations to be considered for straightening, e.g. 15, 249, 1153
growth and polynomial, e.g. 85, 422
linearizable, e.g. 15, 249, 1153
response, comparison of 703, 709, 772, 1041
Curvilinear regression, e.g. 732, 1238
see also nonlinear regression
Cut set method: see minimal . . .
Cutting scores in selection procedures 1007
Cyclic:
data: oval plots 1319
designs 1036
Cyclical data: plotting 1319
Cycling, nutrient, e.g. 1065

d or *δ:* see Somers' delta

Darstellung:
  sprachliche 110
  von Resultaten: eine Einführung
      1151
  wissenschaftlicher Zusam-
      menhänge 328, 379, 1263
Data analysis 8, 69, 70, 123, 200, 351,
    417, 520, 546, 547, 608, 612, 659,
    736, 755, 829, 830, 871, 916, 932,
    996, 1001, 1024, 1137,
    1318–1323, 1382, 1424
  see also BMDP, IDA, SAS, SPSS
      as well as statistical computing
      and microcomputer
  application of pattern recognition
      techniques to 1424
  applied to quantitative summaries
      of different studies: see meta-
      analysis
  empirical Bayes, e.g. 175, 327
  for engineers: a guide to books 480
  geometric 755
  graphical methods, e.g. 69, 70, 183,
      520
    see also graphical displays
  in the presence of error, e.g. 520
  new trends 69, 70, 123, 547, 612,
      755, 830
  for data base design 576
  of two-way and three-way contin-
      gency tables 659, 1320
  multivariate: see multivariate
      methods
  projection pursuit methods for,
      e.g. 755
  spline functions in 1416
Data- and methodbase systems 104,
    507
Data:
  and results, interpretation of, e.g.
      229, 328, 379, 888, 916,
      1151–1153, 1263, 1318–1323
  and their requirements 197, 932,
      1074, 1243, 1284, 1424
  arising in paleo magnetism, navi-
      gation of birds and bees, or-

bits of comets, and sedimen-
    tary geology: see statistics on
    spheres
banks relating to the frequency of
    occurrence of events or acci-
    dents, the reliability of com-
    ponents and the conse-
    quences of accidents, e.g. 524
base:
  design, data analysis for, e.g. 576
  systems 104, 576, 1196, 1330,
      1349
-bases,
  difficulties in numerical compu-
      tation and, e.g. 520
  fuzzy 464
binary, analysis of e.g. 227, 288,
    864
bivariate, methods for investi-
    gating 47
blocks: simplifying complex
    blocks of data: see cluster
    analysis
browsing, data snooping 25, 67,
    160, 788, 1076, 1347
Bundesrepublik Deutschland,
    siehe dort
censored see there
classed in a two-by-two frequency
    table: see fourfold table
    analysis
collection of, e.g. 889
DDR 1255
description 47, 123, 828, 830,
    1317–1322
display:
  see also displaying and displays
  qualitative data analysis, e.g.
      889
distribution of final digits in 1074
dredging, data sniffing, data
    snooping 25, 67, 160, 788,
    1076, 1214, 1347
editing 160, 932, 1074, 1076, 1185,
    1243, 1424
entities, attributes and transactions
    576

fitting equations to 249, 891, 1041, 1376

FRG 161, 162, 163, 281, 584, 591, 967, 1079, 1202, 1255

how much to look at the 25

international outline 226, 1255, 1372

looking at the: see data browsing

matrices and multiway tables: reduction and analysis 47

matrix, reducing a ... to triangular form by using orthogonal transformations, e.g. 661, 824, 1205

missing: see missing data, missing observations

model design 576, 1024

modeling:
  nonparametric statistical 547, 1024, 1025
  using quantile and density-quantile functions, e.g. 547, 1025

multifactor, analysis of: see multi-variate methods

multivariate:
  analysis of: see multivariate methods
  display of 47, 69, 70, 183, 342, 351, 370, 634, 683, 1209, 1317, 1319, 1359, 1366

organisation in data bank systems 1330, 1349

points, erratic, danger of, and breaking down of the theoretical assumptions, e.g. 634

polytomous: generalized linear models, e.g. 864

preprocessing and computational analysis 36, 197, 1243

qualitative:
  analyzing, e.g. 660, 1436
  analyzing: see also contingency table analysis

quality control and editing 36, 160, 197, 888, 932, 1074, 1076, 1185, 1243, 1424

quantitative, sensitive, obtaining 329

reduction:
  see data analysis and cluster analysis
  to a 2×2 table: see fourfold table analysis

re-expression of 502, 831, 916, 1320, 1323

screening 8, 160
  and data transformation 8

sensitive, obtaining, e.g. 311, 329, 820, 1260

sets with missing observations, analysis of 896

simplifying complex blocks of: see e.g. cluster analysis

sniffing,
  data dredging, data snooping 25, 67, 1076, 1347

snooping 25, 67, 160, 1076, 1284, 1347

sources for international data 1372

structures:
  in survey and opinion research, e.g. 1243
  stepwise reduction of 507

summary by a few values of the sample quantile function 547, 1024, 1025

UK 1106

unbalanced, analysis of 789, 1142

United Kingdom 1106

USA 671, 1372
  and the world 1372

Datenanalyse 123, 200, 417, 520, 547, 608, 612, 736, 755, 830, 916, 932, 1001, 1024, 1318–1323
  -Systeme 93, 114, 285, 736, 897, 1195, 1284, 1328
    siehe auch BMDP, IDA, SAS und SPSS
    siehe auch Programmsysteme

Datenbanken 26, 104, 508, 576, 1185, 1196, 1330, 1349

Datenbanksysteme: siehe Data base systems

Daten:
"die Welt in Zahlen" 226, 1255, 1372
Datenerhebung in der Soziologie 117
Datengewinnung:
Bevölkerungsstatistik 161, 340
Wirtschafts- und Sozialstatistik 727
Daten:
internationale Übersichten 226, 1255, 1372
medizinische: Zusammenhangsanalysen, z. B. 1384
Datenmodellierung: Theorie und Praxis des Datenbankentwurfes unter besonderer Berücksichtigung des relationalen Modelles und seiner Weiterentwicklungen 576
Datenquellen für Sozialmedizin und Epidemiologie 133
Daten:
zensierte 134, 761, 762, 770, 893, 1177
siehe auch censored survival data
zu den Themen Mensch, Ernährung, Rohstoffe u. a. m. 226
zur Bundesrepublik Deutschland 161, 162, 163, 281, 415, 539, 553, 584, 591, 967, 1079, 1202, 1255
DDR: Daten 1255
Death rates: see mortality analysis
Decision:
analysis and decision-aiding models 398, 654, 1119, 1120
analysis:
see also decision theory
and fuzzy sets 464, 1445
in insurance 528, 575, 1300
in the pharmaceutical industry 1301
processes and fuzzy sets 464, 1445
rules for allocating treatments in sequential medical trials 55
theory 48, 66, 87, 324, 398, 498, 528, 536, 654, 757, 758, 769, 880, 1053, 1119, 1120

and Bayesian methods, elementary, e. g. 66
-tree and problem solving 192
Decomposing chi-square 103, 263, 352, 793, 836, 1100, 1110, 1153, 1334
Decomposition of matrix transformations: eigenstructures and quadratic forms, e. g. 445
Deduktion und Induktion 182, 684
Defective items: effects of false and incomplete identification on the reliability of acceptance sampling 700
Defectives: too many in too short a time? 939
Defective units in a lot: confidence limits 984
Defense, systems analysis and modeling in 581
Delphi:
Studien 180, 247
studies 180, 247
Delta, Somers' delta 9, 611, 766
Deming-Stephan algorithm, e. g. 472
Demodulation, complex 718
Demographic:
ageing, components of 818
data, problems of, e. g. 750
processes, e. g. 193, 601
trends: see Demographie and Bevölkerungsstruktur
Demographie 193, 340, 601, 667, 750, 818, 851, 1148, 1232, 1281, 1404, 1428
siehe auch Bevölkerungswissenschaft
bibliographische Hinweise 340
anderer Länder 226, 671, 1106, 1148, 1255, 1372
Anwendung neuerer Zeitreihenverfahren 626
der Bundesrepublik Deutschland 161, 162, 163, 281, 415, 539, 553, 584, 591, 967, 1079, 1148, 1202, 1255

28

Demographische:
Prozesse in der Bundesrepublik Deutschland und das System ihrer statistischen Beobachtung 161
Variablen: Systemzusammenhänge einschließlich von Rückkopplungen, z. B. 505
Demography 193, 340, 667, 750, 818, 851, 1148, 1232, 1281, 1404, 1428
deterministic aspects in mathematical 601
Frobenius and Leslie matrices in, e.g. 54
historical, sources and methods of, e.g. 1404
multidimensional mathematical 750
of ageing in Britain and the USA 818
the discrete time recurrence model, the continuous time model, and the discrete time matrix model 601
Demometrie: siehe Bevölkerungsstatistik
Demoskopie 967
siehe survey and opinion research
Dendrogram: see cluster analysis
Density estimation 47, 63, 401, 658, 719, 1042, 1071, 1289, 1345, 1386
for multivariate data, e.g. 1071
for univariate and bivariate data, e.g. 47, 1071
nonparametric:
developments in 63, 1024, 1071
probability 1289
Dependence, joint dependence, e.g. 1323
Descriptive statistics: see e.g. 701
Design:
balanced and unbalanced in analysis of variance, e.g. 1259
experimental:
see experimental design
and non-experimental for social research, e.g. 117, 220, 1246, 1404

information function, e.g. 120
Designing experiments:
discriminating among separate models 1046
Design:
of block experiments 1257
of experiments:
see also experimental design
developments in the 33, 35, 477, 1257
for mixture problems 1257
of factorial experiments 1257
of mixture experiments 1257
Designs:
for nonlinear models, e.g. 1257
neighbour 40
Deskriptive Statistik, z. B. 43, 81, 349, 1374
Desktop computation 106
see microcomputers and statistical computation
Destructive (nonrepairable) failures of single units, e.g. 974
Detection:
limits in analytical chemistry 244, 320, 577, 1280
or exploratory sampling 1395
Detective work and explorative statistics, e.g. 871, 1318–1322
Determinants, e.g. 54, 749, 1207
and matrices: see matrix algebra
DETMAX, computer algorithm to find D-optimum designs 1257
Deutsche:
Industrie Normen für Größen und Einheiten in Naturwissenschaft und Technik 279
Sprache 110
Developmental models 954
Diagnosen und Normabweichung klinischer Daten 390
Diagnostic:
checking in forecasting, e.g. 4, 122
plots 249, 547
for explanatory variables 34
regression analysis, graphical methods of 34

Diagnostic:
  screening, indices for 79, 367
  sensitivity and specifity 2, 78, 79,
    86, 213, 250, 367, 390, 620,
    682, 986, 1355
Diagnostics:
  Pregibon's, for generalized linear
    models illustrated with an
    example on logistic regression
    219
  regression 75, 219, 442, 550, 755,
    1043, 1077, 1264, 1340, 1383
Diagnostic tools for data analyses,
  e.g. 219
Diagnostik:
  Methoden der Statistik und Infor-
    matik in der 78, 79, 86, 132,
    250, 390, 986, 1355
  und Prognostik mit Tumormarkern
    2, 986, 1355
Diagnostische Tests 78, 79, 86, 213,
    250, 367, 390, 682, 986, 1355
Diagonal elements of the hat matrix
    219, 442, 547, 1264
Dichtemittel und Isoplethen 1155
Dichtefunktionen: ihre Schätzung 47,
    63, 183, 401, 1024, 1071, 1289,
    1345, 1386
Dichte- und Hazardfunktionen bei
    zensierten Daten, nichtparame-
    trische Schätzung von 692
Dictionary:
  and classified bibliography of sta-
    tistical distributions in scien-
    tific work 1030
  of statistical terms 657, 991
Differential:
  equations, e.g. 56, 1085
  calculus and matrix algebra: ele-
    mentary hints 798
Differential- und Integralrechnung
    sowie gewöhnliche Differential-
    gleichungen, z.B. 56, 1085
Diffusion processes, e.g. 50, 51, 53,
    454, 966
Digital computers, statistical
    methods for, e.g. 1085

  see also statistical computing and
    Programmsysteme
Dimensionality:
  and misclassification rate 490
  reduction of 717
Dimension reduction and orientation
  of multivariate observations,
  e.g. 1209
DIN: Normen für Größen und Ein-
    heiten in Naturwissenschaft und
    Technik 279
Directional data statistics 57, 364,
    719, 839, 840, 1029, 1373
Directly adjusted rates 103, 194, 380
Disaster planning: preservation of
    life and property 375
Discrete:
  control schemes, design of, e.g. 122
  distributions 298, 504, 542, 631,
    632, 701, 1028
    developments in 632
    mixtures of 343
    relationships between 542
  maximum principle applied to
    optimal systems reliability 1308
  random variables, e.g. 454, 643
  replacement models 999
  -state stochastic models in commu-
    nication processes 900
  variates, e.g. 454, 643
Discriminant:
  (function) analysis 8, 12, 31, 62, 83,
    93, 114, 285, 330, 346, 370,
    446, 482, 490, 501, 513, 537,
    583, 612, 634, 680, 689, 730,
    740–743, 767, 798, 800, 835,
    840, 869, 870, 952, 972, 1085,
    1195, 1209, 1240, 1284, 1350, 1374
  analysis for time series 717
  analysis, selection of variables in,
    e.g. 717
Discriminating among separate
    models 165, 1046
Discrimination:
  and allocation with discrete data,
    e.g. 537
  and classification 537, 634, 658,
    1078

application of matrices in, e.g. 54

models, assessment of Fisher and logistic linear and quadratic 62

Disease:
chronic:
disease risk, longitudinal models for, an evaluation of logistic multiple regression and alternatives 1421
screening in 910
complex interactions with covariates, Poisson regression models for evaluating 386
early detection and screening procedures 419
incidence:
logistic regression 813
prevalence and mortality measures, e.g. 682
incubation period, median 27
interaction contrast disease rates for assessing synergism or antagonism in multifactor-multilevel disease risks 1091
rare: two group trial sample sizes 780
rate modeling with Poisson regression analysis 386
rates for assessing synergism or antagonism in multifactor-multilevel disease risks 1091
risk 84, 137, 338, 367, 418, 538, 682, 1021, 1091, 1182, 1231, 1363, 1364, 1429
statistics, mathematical models for infectious 35
the spread of a, e.g. 1065
time-space clustering of 225, 682
Disequilibrium, relative, as a measure of interaction in contingency tables 659
Diskriminanzanalyse 8, 12, 31, 62, 83, 93, 114, 285, 330, 346, 370, 446, 482, 490, 501, 513, 537, 583, 612, 634, 680, 689, 730, 740–743, 767, 798, 800, 835, 840, 869, 870, 952, 972, 1085, 1195, 1209, 1240, 1284, 1350, 1374
und Clusteranalyse: Beispiele, Methoden und Programm-pakete 262
Diskriminationsverfahren 330, 490, 537, 634, 869, 1046
Dispersal, density-independent and density-dependent, e.g. 1065
Dispersion:
and location, measures of, e.g. 658
in categorical data 1170
in paired samples, e.g. 1162
Displaying:
multivariate data 47, 69, 70, 183, 342, 351, 370, 634, 683, 1209, 1317, 1319, 1359, 1366
one-, and twodimensional data: see graphische Darstellungen und graphische Methoden
Displays:
residual, the robustness of 756
semigraphic: see box-plot variations and stem-and-leaf displays
Dissimilarity, measuring: see cluster analysis
Dissoziationskonstanten, z.B. 1286
Distance:
and similarity: measurement of as well as graphical display of . . . matrices 342
measures for distributions and estimators 406
similarity and dissimilarity: see cluster analysis
Distributed data base systems 61, 1330
Distribution:
see also Verteilung
a reliability . . . with increasing, constant and bathtub-shaped failure rates 544
Distributionfree procedures 214, 312, 363, 369, 703, 719, 770, 793, 836, 893, 1024, 1072, 1090, 1225, 1288

Distribution:
  of goals per game scored by the
    England soccer team
    (1872–1983) and fitted
    Poisson and negative
    binomial distributions 1063
  of justice sentiments experienced
    by members of a social
    aggregate 613
  of the time of a strike to be settled,
    of the length of stay of a
    patient in a hospital, and of
    the time of a disease to lead to
    recovery or dead 763
Distributions:
  see also statistical distributions
  see also under proper name
  and inference problems on the
    sphere, e.g. 839
  bounded binomial and multi-
    nomial 747, 748
  Burr and related 541, 1285
  characterization of, e.g. 454, 1028
  circular distribution and distribu-
    tions on the sphere 57, 364,
    839, 840, 1029, 1373
  contagious, e.g. 298
  continuous 504, 541, 631, 701, 760,
    1028, 1030
    multivariate, e.g. 631, 1030
  dictionary and classified biblio-
    graphy of statistical 1030
  discrete 103, 298, 504, 542, 631,
    632, 701, 760, 1028, 1030
    developments in 632
    generalized power series, e.g.
      542
    of order k, e.g. 542
  finite mixture 343
  longtailed, understanding some
    1127
  mixtures of 307, 343, 869
  multivariate: see there
  noncentral: see noncentral distri-
    butions
  number-of-errors, equality of, dis-
    tances between and orderings
    of 659
  of final digits in data 1074
  of quality and quantity goods 613
  Pareto 30, 504, 541, 613, 631, 1028
  properties of 20 continuous and
    5 discrete 1028
  relations between discrete 542
  shapes of 1320
  standard, e.g. 658
  statistical, e.g. 454, 504, 541, 542,
    631, 658, 701, 768, 769, 1028,
    1320
    dictionary and classified
      bibliography 1030
  systems of 504, 541, 542, 631, 658,
    701, 768, 769, 1028, 1030, 1320
  three-parameter: lognormal, gam-
    ma and Weibull, on the
    parameter of the, e.g. 735
  useful in general insurance work
    (at an elementary level), e.g.
    575
Distributive justice, a new theory of
  613
Divided schedule and/versus imputation for nonresponse, e.g. 248
Divisia index numbers 614
Documentation of clinical trials 96,
  160
Documenting the results of a study
  160
Dokumentation 399, 746
  wissenschaftlicher Studien 160
Dominance in marital interactions,
  tests of 1365
Dominanztest für sequentiell-katego-
  rische Daten 1365
Dorfman-type group testing, exten-
  sions of 592, 812, 923, 1051
Dose response studies 160, 173, 359,
  360, 531, 582, 1095, 1332
Dosis-Wirkungs-Analysen, qualita-
  tive, Einzelsubstanzen und
  Kombinationen 1332
Dot charts without or with grouping,
  e.g. 200
Double sampling, e.g. 621, 1181
Dreiecksform der Input-Output-
  Matrix, e.g. 456

Dreieckskoordinaten 56, 915
Dreieckstest 1393
Drolette's estimator, e.g. 194
Drug:
    evaluation: role of statistics, e.g.
        160, 1353
    quality control of the manu-
        facturing process 160
    research, e.g. 88, 160, 1353
    screening, e.g. 160
Drugs:
    new drugs: efficacy and safety
        evaluation 160, 1353
    shelf life 160
    stability of 160
Dual scaling:
    and contingency tables 965
    and correspondence analysis 965
    internal consistency of categorial
        data 965
    of categorical data 965
Dummy (indicator) variables in re-
    gression analysis 4, 300, 425,
    564, 681, 798, 955, 1383, 1437
Duncan-Test: Schranken 921
Dunnett's procedure 59, 71, 309, 787,
    1067, 1187
    Shaffer's extension to 787, 1067
Dunn-Test: verbesserte Schranken
    391
Durbin-Watson Test 4, 300, 672, 673,
    1350, 1437
Durchschnittliche Wachstumsrate in
    der Wirtschaft 20, 268, 1171
Dynamic:
    programming 1308
    systems, linear and nonlinear:
        Bayesian methods 143
Dynamische Optimierung 695

Earnings equations, using longitudinal
    data to estimate age, period and
    cohort effects in 851
Eckart-Young decomposition, e.g. 54
Ecological statistics, e.g. 222, 223,

995, 996, 997, 1031, 1065, 1113,
    1208
Ecology:
    application of canonical analysis
        411
    data analysis 996
    multivariate methods 997
    quantitative, and related
        ecometrics 1031
    sampling biological populations
        223
    spatial and temporal analysis in
        222
Econometric:
    analysis of economic time series,
        e.g. 523
    applications:
        of discrete data structural analy-
            sis 833
        of loglinear models 355
Econometrics, e.g. 650, 840
    and quantitative ecology 1031
Economic analyses: optimal triangu-
    lation of the input-output tables
    of the FRG and the European
    Community 456
Economics: a game theoretic ap-
    proach, e.g. 128
Economic:
    statistics e.g. 913
        for comparative assessments 35
    time series, analysis of 523, 650,
        777, 953, 1265
EDA:
    Exploratory Data Analysis 47, 109,
        197, 229, 417, 435, 502, 547,
        596, 608, 659, 755, 828, 829,
        830, 871, 915, 916, 965, 1024,
        1318–1323, 1339, 1347, 1424,
        1436
    computer programs:
        written in APL and FORTRAN
            871
        written in BASIC and
            FORTRAN 1339
ED_{50}-Bestimmung nach Litchfield
    und Wilcoxon, z.B. 1286

Editing data 160, 932, 1074, 1076, 1185, 1243, 1424
EDV: Einsatz in Erhebungen, z. B. 1243, 1424
Effective communication 328, 379, 1263
Effect:
  size:
    computation of, e. g. 414
    in social research, e. g. 414, 588
    percentile equivalents of, e. g. 414
  sizes:
    cumulation formulas for meta-analytic methods 414, 588
    transforming significance levels into, e. g. 414
Ehestatistik 161, 591, 1079, 1255
Eigenvalues:
  and eigenvectors, e. g. 1207
  norms and bounds for, e. g. 749
Einflußkurve: siehe influence curve
Einkommen, Verdienste: Daten 1079, 1255
Einstichprobentest für den Vergleich einer empirischen mit einer theoretischen Varianz 802
Einstichproben-T^2-Test nach Hotelling 800
Einzelfallbetrachtung: Veränderungsmessung 117, 413, 527, 571, 1047, 1348
Elderly people:
  in the populations of developed societies 818
  statistics of 161, 340, 539, 818, 851, 1148, 1255
Electricity demands forecasting 432
Elektronik: Zuverlässigkeitsanalyse, z. B. 1168
Elektrotechnik und Maschinenbau: Beispiele zur Zuverlässigkeitsanalyse 1130
Elementare (einfache) statistische Methoden 121, 1151–1153, 1238, 1244, 1318–1323, 1351
ELIM, elimination algorithm for calculating estimates and standard errors (regression analysis), e. g. 15
Ellipses: simultaneous confidence ellipses for pairs of mean components, e. g. 634
Ellipsoids: simultaneous confidence intervals and ellipses as shadows of the p-dimensional ellipsoids, e. g. 634
Elveback's estimator, e. g. 194
EM algorithm 634, 804, 805
  survival data analysis 234
Emigration and immigration processes, e. g. 193
Empirical:
  Bayes data analysis, e. g. 175, 327
  cumulative distribution function, e. g. 640, 643, 658, 719, 1024, 1042, 1052, 1072
  distribution function, e. g. 640, 643, 658, 719, 1024, 1042, 1052, 1072
  tests, e. g. 1052
  influence curve, e. g. 219
Empirische Studien, Auswertungskonzepte für, z. B. 1347
Encyclopedia of statistical sciences 701
End-point rule, e. g. 871
Endziffern: sind sie gleichverteilt? 1074
Energy:
  demands forecasting 269, 432
  flow and nutrient cycling through an ecosystem, e. g. 1065
  statistics, e. g. 432
Engineering systems, reliability evaluation of: see Reliability of engineering systems
Engineer's guide to books on statistics and data analysis 480
Ensemble-adjusted p values 1133
Entity-relationship modelling 576, 1330, 1349
Entropie als Maß in der Bevölkerungsstatistik 692

Entropy, statistical inference and measurement of 123

Entscheidungstheorie 48, 87, 324, 398, 498, 528, 536, 654, 757, 758, 769, 880, 1053, 1119, 1120

Enumeration:
partial and Geoffrion implicit 1308
Polya's formula 1315

Enumerative survey sampling methods, e.g. 820

Environmental:
exposures:
and cancer 292, 338, 888, 1021
and disease risks: see there
statistics for biologists 447

Enzyklopädie der Statistik 701

Enzym-Bestimmungen im Vergleich, z.B. 172

Enzymkinetik 1286

Epidemic models for the diffusion of e.g. news and rumors 50, 51, 53, 966

Epidemiologic:
methods 6, 35, 86, 162, 386, 418, 450, 682, 738, 742, 753, 797, 1139, 1140, 1230, 1429
modeling: Reed-Frost models 171
terms, e.g. 682, 753

Epidemiology:
application of discriminant analysis in 742
associations in 21, 338, 682, 1429
dictionary of 753
pitfalls and fallacies, e.g. 450, 1230, 1429
weak associations in 1429

Equations:
numerical solution of 54, 443, 1085, 1207
systems of:
least absolute value estimation in 275
small samples and large 720

Equilibrium:
laws in panmictic populations 331
states and time to equilibrium, e.g. 420

Equipartite designs 1036

Equivariance, e.g. 774

Ergodic theorem, e.g. 454

Erhebungen 36, 176, 196, 205, 248, 278, 329, 367, 373, 422, 558, 564, 594, 621, 644, 676, 727, 804, 820, 845, 872, 992, 1137, 1161, 1200, 1243, 1262, 1274, 1275, 1276, 1342, 1371, 1382, 1403, 1433

Erhebungsbogen, z.B. 609, 776, 1243

Erlang distribution 504, 552, 631, 760

Erneuerungstheorie, z.B. 32, 420, 454, 695, 1136, 1304

Error:
distributions of errors 235, 733
estimation and theory 494, 624, 726, 733, 880, 1102, 1342, 1343, 1435
propagation of error, formulas 733

Errors:
in economic statistics, e.g. 913
in particulate material sampling 469
in population data, e.g. 166
in surveys: see surveys
type III, and multiple comparisons 111

Erwerbspersonen und Erwerbsstatus: Daten 161, 591, 1079, 1255

Erwerbstätige pro Wirtschaftszweig: Daten 161

Erwerbstätigkeit:
und Arbeitslosigkeit: Daten 161, 591, 1079, 1255
und Arbeitsmarkt: Daten 20, 591, 1079, 1255

Erzeugung:
und Prüfung von Zufallszahlen, z.B. 1190
von Zufallsvariablen 421, 1035, 1117

Escalation, structural: see there

ESD (extreme Studentized deviate) many-outlier procedure, percentage points for a generalized 1135

Estimating:
  log-linear models, e.g. 1110
  the size of a closed population 103,
      966, 1000, 1064, 1065, 1208,
      1210, 1391
Estimation:
  see also estimation theory
  interval, e.g. 66, 658, 769
  inverse median 1324
  methods and relationships be-
      tween them, e.g. 66, 490
  of parameters in restricted multiple
      regression and loglinear
      models 1115
  of the probability of an event as a
      function of several indepen-
      dent variables 1360
  point, e.g. 66, 774, 1028, 1090
  robust:
    a density-quantile function
        perspective on 756
    and robust estimation in the
        presence of outliers 756
    Pitman-like 756
  theory, e.g. 66, 580, 658, 769, 774,
      880, 1071, 1090
  unbiased, e.g. 774
Estimators:
  desirable properties, e.g. 490
  L-, M-, R- and W-estimators, e.g.
      547, 719
  minimum L_P norm, e.g. 661
  of location: a comparison 547, 916,
      1247, 1248
  robust 253, 488, 547, 580, 587, 828,
      1209, 1247, 1248
Evaluation, multiattribute 324
Evans county heart disease study,
    e.g. 682
Even-resolution designs, e.g. 538
Event:
  how long is the wait for an uncom-
      mon event? 1145
  trees and fault trees, e.g. 99, 378
Evidence: convincing evidence for a
    judge 299, 354, 356, 959, 1241
Exakter Fisher-Test:
  für eine $r \times c$-Tabelle 877

  für eine $2 \times k$-Tabelle 653
Excess life: see renewal process
Expectation:
  -maximization (EM) algorithm:
    and some illustrations: expo-
        nential distribution, gamma
        distribution, and multi-
        nomial distribution 234
    nonresponse in sample surveys
        804, 805
  of random variables, e.g. 454
Expected:
  mean squares, computing, e.g. 661
  normal scores transformation 943
Experimental:
  and control condition (integrating
      studies): the probability of
      failing to detect an effect size
      414
  and non-experimental designs in
      the social sciences: an intro-
      duction 1246
  design 33, 35, 121, 123, 251, 270,
      348, 413, 474, 477, 517, 525,
      538, 578, 579, 658, 674, 769,
      896, 955, 1036, 1076, 1093,
      1096, 1107, 1122, 1246, 1257
    and regression, the interrelation-
        ship between, e.g. 83
    bibliographies 33, 348, 477, 1257
    developments in 33, 348, 477,
        1257
    interactions 131, 232, 417, 525,
        674, 716, 789, 1036, 1122,
        1390
    John's three-quarter fractional
        factorials 270
  designs:
    advantages and disadvantages
        of the main designs used in
        the behavioral sciences 674
    "error-robust", "model-robust",
        and "model-sensitive" de-
        signs 1257
    for non-linear models 1257
    special classes 1036
Experimente: welche Mängel sind zu
    beklagen? 1076

Experiments:
  designed experiments, regression
    models and calculations, e. g.
    824
  design of comparative, e. g. 769
  iterative cycles of investigation 121
  resulting in unbalanced data and
    their analysis 789, 1142
  spoilt experiments 1036
  two-drug chemotherapy,
    describing the response
    surface in exploratory 1272
  with mixtures: designs, models
    and the analysis of mixture
    data 224, 333, 1257
Expertensysteme, z. B. 153, 256, 478,
    1163, 1346
  und Knowledge Base Manage-
    ment Systems 104, 1108
Expert:
  guidance embedded in statistical
    programs 478
  systems, e. g. 153, 256, 478, 1163,
    1346
    and expert knowledge in statisti-
      cal computing 478
  witness, the role of the statistician
    as an 299, 356, 959
Explanation and models: see models,
    and simulation
Explanatory variables and survival
    data 234
Explorative Datenanalyse 47, 109,
    229, 417, 435, 502, 547, 596, 608,
    659, 755, 828, 830, 871, 915, 916,
    965, 1024, 1318–1323, 1339,
    1347, 1424, 1436
Exploratory:
  analysis of contingency tables 659
  data analysis 47, 109, 229, 417, 435,
    502, 547, 596, 608, 659, 755,
    828, 830, 871, 915, 916, 965,
    1024, 1318–1323, 1339, 1347,
    1424, 1436
Exponential:
  distribution 504, 541, 631, 701, 770,
    847, 952, 1186

and other continuous distribu-
    tions, mixtures of 343
  generalized, e. g. 420
  growth equation, e. g. 1065
Exponentiality, tests for 719
Exponential smoothing, e. g. 4, 186,
    1019
  double and triple, e. g. 4
Exponentielle Glättung, z. B. 4, 185,
    186
Exposure:
  and disease, e. g. 84, 338, 1021,
    1231, 1429
    factor as well as bias in
      estimating exposure 1182
  and risk
    see also disease and risk
    at least 2 levels of exposure and
      methods of analysis 137
  to hazardous material: risk
    assessment 1204
External and internal scaling (regres-
    sion diagnostics) 219
Extreme:
  Studentized deviate many-outlier
    procedure, percentage points
    for a generalized 1135
  vertices designs 1257
Extremwert-Verteilung 504, 631, 952
Eyring equation 160

Faces: see Chernoff-type faces
Factor analysis 8, 22, 93, 114, 221,
    285, 344, 358, 370, 417, 482, 501,
    603, 604, 608, 612, 628, 634, 661,
    681, 798, 827, 840, 855, 900, 972,
    1078, 1085, 1109, 1195, 1374
  confirmatory, e. g. 806, 807
  confirmatory, conceptual replica-
    tion and, e. g. 588
  in chemistry 827
Factor and cluster analysis, e. g. 900
  FORTRAN programs, e. g. 608
Factorial:
  and split-plot designs,
    asymmetrical 251

Factorial:

designs 121, 251, 322, 438, 463, 477, 525, 674, 955, 1036, 1073, 1084, 1107, 1122, 1234, 1238, 1257, 1408

  fractional designs with treatment-interaction confounding 674

  interactions and confounding in 1036

  mathematical theory of 1084

  nonorthogonal, e.g. 83

  replicated, based on alignment: tests of interaction 543

  staggered nested, use and analysis of 1234

  with group-interaction confounding 674

experiments:

  see also factoral design

  and marginality 1073

  multiple comparisons for orthogonal contrasts 68

Factor:

scores, estimation of, e.g. 1078

selection by MAICE criterion, e.g. 1078

Failing systems: see maintenance schedules and reliability analysis

Failure:

data: see reliability of engineering systems

distributions, multivariate, e.g. 974

mode and effects analysis and fault tree analyses 378

probability confidence limits, e.g. 99

process, generalised, for non-maintained systems 378

rate:

  decreasing: mixed exponential method applied to reliability, e.g. 265

  estimation by nonparametric methods, e.g. 1226

  instantaneous: see hazard rate function

  rates 21, 99, 420, 544, 770, 951, 1226

-repair process, e.g. 524

time:

  model, inference based on ranks in the accelerated, e.g. 642

  models 32, 234, 560, 642, 770, 847, 950, 952

  to first ... for new systems, e.g. 420

  types in survival data analysis 234, 332

Faktorenanalyse 8, 93, 114, 221, 344, 358, 370, 417, 482, 501, 603, 604, 608, 612, 628, 634, 661, 681, 798, 827, 840, 855, 900, 972, 1078, 1085, 1109, 1195, 1374

Anwendungen in der Chemie 827

Faktorielle Versuchspläne 121, 322, 438, 463, 477, 525, 674, 955, 1036, 1073, 1084, 1107, 1122, 1234, 1238, 1408

siehe auch factorial designs

Yates-Algorithmus 438, 463

Fallacies in the analysis of subgroups, e.g. 588

Fall-Kontroll-Studien 6, 86, 134, 137, 142, 225, 258, 338, 367, 561, 791, 813, 1182, 1230

Falsifizierung von Hypothesen 182, 291, 1066, 1401

Falsifying probability statements 291

Familienstandsgliederung 161, 1079, 1255

Family life cycle, e.g. 194

Fault:

tolerant computers and fault software, reliability of, e.g. 46

tree analyses, e.g. 99, 378, 524

Faulty inspection distributions, some generalizations, e.g. 265

F distribution:

algorithms and tables 841

and its limiting distributions 1153

exact p values for F tests 744

generalized 560

noncentral 462, 489, 525, 541, 970, 1028, 1032

percentage points 661, 841
probabilities 661, 841
tables of the ... and related distributions with algorithms 841
Fecundability 281, 340, 626, 1148, 1428
Federal data collection: pressures for more detail, greater frequency, and more prompt availability 846
Feeling-of-knowing predictions, measures of the accuracy of 947
Fehlende Werte: Behandlung in BMDP, SAS und SPSSx 692
Fehlerabschätzung und Fehlertheorie 494, 726, 880, 1102, 1342, 1343, 1435
Fehlerbaum-Methode, z. B. 99, 378, 524, 1168
Fertilitätsstatistik: 340, 818, 851, 1428
Fertility:
    schedules, marital, analysis of age, period and cohort effects in time series of age-specific 851
    table, e. g. 194
    total fertility rate, TFR, cohort TFR and period TFR 851
    trends 340, 818, 851, 1428
Fiducial argument, fiducial inference, e. g. 48, 658, 643
Field experiments 1036, 1247
Filliben test, e. g. 808
Filtering:
    adaptive, e. g. 811
    harmonizable filtering and sampling of time series, e. g. 720
Finanzen, öffentliche Haushalte: Daten 1079, 1255
Finanzmathematische Tabellen 246
First:
    -order and zero-order models 160
    passage time distribution 39, 184, 372, 454
Fisher's exact permutation test, Fisher's exact test, Fisher's test 766, 793, 836, 1072, 1152, 1153, 1335, 1434

Fisher's exact test 766, 793, 836, 1072, 1152, 1153, 1335, 1434
    see also fourfold table analysis
    computation of two-tailed 1293
    continuity correction for 1003
    extension to $r \times c$ contingency tables 877, 1012
    extension to $2 \times r$ contingency tables 653
    F-test alternatives to 1004
    in $r \times c$ contingency tables 877, 1012
Fisher's test on the significance in harmonic analysis, a Monte Carlo program for the critical values of 1144
Fisher-Test, exakter 766, 793, 836, 1152, 1153, 1335, 1434
    Verallgemeinerungen 653, 877, 1012
    und Somers' Delta, z. B. 766
    zweiseitiger 1293
Fish patties, using three fish species and three processing factors, preparing 224
Fit, tests of: see goodness of fit
Fitting:
    a general log-linear model, an algorithm for 471
    data 871, 916, 1098, 1320
    equations:
        to data 249, 891, 1041, 1098, 1376
        to multifactor data 249
        to nested data 249
    models to survey data: see survey
    straight lines when both variables are subject to error 147, 300, 547, 580, 587, 641, 832
    unbalanced two-way data by additive models 1142
Fixed effects:
    analysis of variance, e. g. 362, 412
    analysis of variance and covariance, consequences of failure to meet assumptions underlying the 412, 661

Flächen-Stichproben 621, 1061
Flexible sampling 1395
Floating-point computation, error in, e.g. 661
Floods, statistical analysis of 35
F_{max}-test, a note on the critical values of the 652
F_{max}-Test: Hinweis zu den kritischen Schranken 652
Folgetestpläne 694, 769, 1348, 1374, 1388, 1393
Follow-up studies:
  see also cohort studies
  coronary heart disease, e.g. 682, 1314
  medical, e.g. 96, 194, 682, 694, 1314, 1348
Forecasting 4, 107, 122, 150, 185, 186, 199, 261, 526, 557, 784, 785, 825, 826, 881, 985, 1019, 1020
  accuracy of, e.g. 825
  crop 35
  diagnostic checking in exponential smoothing, regression models, time series, transfer function-noise models, e.g. 4
  Holt-Winters methods with listed computer programs, e.g. 4
  long term, e.g. 150
  short term, e.g. 4, 985
  through data analysis, e.g. 199, 784, 785
Forecasts: see forecasting and Box-Jenkins forecasting
Form, determination of, e.g. 1113
Formaldehyde carcinogenicity in rats and mice, e.g. 198
Formelsammlung und Definitionen zur beschreibenden und schließenden Statistik, z.B. 1352
Formulierung von Texten statistischen Inhalts 328, 379
Forschung und ihre Planung, z.B. 1246
Forst-Biometrie, z.B. 616
FORTRAN programs, e.g. 83, 93, 194, 285, 404, 608, 639, 736, 786, 971, 972, 1098, 1195, 1245, 1249, 1250, 1264, 1284, 1294, 1339
Fourfold:
  chi-square test with $(n-1)$ instead of n in the numerator 1111, 1151–1153, 1335
  table:
    analysis 103, 210, 227, 284, 352, 367, 470, 472, 643, 659, 739, 793, 798, 836, 854, 884, 984, 1002–1004, 1056, 1072, 1110, 1111, 1122, 1151–1153, 1334, 1335, 1337, 1434
    chi-square test of homogeneity and F-test alternatives 1004
  tables:
    combining evidence from 137, 227, 367, 884, 1152, 1153
    combining evidence from: see also Mantel-Haenszel summary odds ratio
    extensions, see contingency table analysis and homogeneity of proportions or distributions
    power of chi-square tests for 1002
Fourier transforms, e.g. 718
Fractals, martingales, and random fields, e.g. 35
Fractiles, 0.025 and 0.975 and their 0.90 confidence intervals: nonparametric and parametric methods, e.g. 441
Fragebogen 36, 125, 167, 609, 1137, 1200, 1276
  und Interview 36, 125, 564, 1276
Fragen, wie sie in Erhebungen formuliert werden sollten 36, 125, 167, 564, 609, 644, 1137, 1200, 1276, 1424
Frage- und andere sehr elementare Arbeitstechniken, z.B. 1444
Framingham study: coronary heart disease risk 1314, 1360

Freeman's theta, e. g. 9, 11
Freizeitbeschäftigungen: Daten 584,
    1079
Frequency:
    distributions 658
    domain models 718, 1080
FRG: Data 161, 162, 163, 281, 415,
    539, 553, 584, 591, 967, 1079,
    1202, 1255
Friedman's rank statistic: generaliza-
    tions 509, 579, 725, 772, 836
Friedman:
    statistic:
        see also Friedman-Test
        probabilities for the 796
    -Test 285, 709, 793, 836, 1152, 1153
        Approximationen 599
    test:
        extended tables 978
        generalizations of the 509, 579,
            725, 772, 836
        in balanced and unbalanced
            block designs 1229
    -Test:
        kritische Schranken 978
        lineare Kontraste 787
        Verallgemeinerungen 579, 725,
            772, 836
        Wahrscheinlichkeiten 796
Fruchtbarkeit:
    Analyse und Prognose für die Bun-
        desrepublik 281, 626, 1148
    Baby-Boom-Generation und nach-
        folgendes Nullwachstum,
        z. B. 505
Fruchtbarkeitsziffer, Entwicklung
    der allgemeinen 626
Fruit punch mixture by blending pro-
    portions of orange juice, pine-
    apple juice, and grapefruit juice
    224
F test: exact p values 744
F tests, combining independent,
    Bayes procedures for 650
Functional estimation, non-
    parametric 1071
Funktionen, mathematische: Bedeu-
    tung in der Biologie, z. B. 56

Functions:
    definition and properties of some:
        concave, convex, (incom-
        plete) beta and gamma, e. g.
        1028
    implicit, e. g. 85
Funktionssicherheit von Anlagen,
    Geräten und Maschinen: siehe
    Zuverlässigkeit technischer
    Systeme
Future energy requirements 432
Fuzzy:
    clustering algorithms, cluster
        validity of 612
    information and decision
        processes, e. g. 464, 1445
    probability, e. g. 1367
    programming, e. g. 1367
    reasoning 464, 1367, 1439, 1445
    sets 464, 879, 1367, 1439, 1445
    statistics, e. g. 1367
F-Verteilungen:
    nichtzentrale 462, 489, 525, 541,
        970, 1032
    zentrale: Tafeln und Algorithmen
        841

Gain scores, e. g. 414
Galton-Watson process, e. g. 53,
    874
Gambling and martingales,
    e. g. 454
Games:
    and puzzles: theories for winning
        games and solving puzzles 89
    with graphs, e. g. 1315
Game:
    theoretic models and applications:
        in economics, e. g. 128
        in political science, e. g. 128
    theory 52, 89, 128, 212, 536, 637,
        758, 1009, 1011, 1092, 1099,
        1302
Gamma:
    (measure of association) 9, 81, 349,
        429, 724, 766, 836, 947

Gamma:
  distribution 14, 99, 119, 234, 304,
    343, 406, 475, 504, 511, 541, 631,
    661, 712, 735, 761, 770, 816, 847,
    864, 1028, 1186
  a two-sample test 1222
  properties of estimators and
    tabulation of a two para-
    meter gamma density 119
  simultaneous estimation in the
    3-parameter case 119
  distributions, comparison of two,
    e.g. 770
  function: see functions
  population, selecting the best 406
  probability plot, e.g. 770
  tail probabilities: a normal
    approximation 14
  -Verteilung 14, 119, 234, 304, 343,
    406, 475, 504, 511, 541, 631, 712,
    735, 761, 816, 847, 864, 1028
  siehe auch gamma distribution
Gart's exact test for comparing
  matched proportions in cross-
  over designs, e.g. 766
Gas-liquid chromatography:
  application of factor analysis 827
Gauss distribution or Gaussian
  distribution: see normal
  distribution
Gaussian:
  elimination and the sweep
    operator, e.g. 661
  process, e.g. 372, 454, 874
  quadrature, e.g. 661
  shape, tests for departures from,
    e.g. 129, 290, 547, 716, 806,
    853, 1152, 1153, 1390
Gaus-Jordan methods, e.g. 661, 890
Gauss-Markov theorem, e.g. 271,
  634, 732, 1420
Gauss-Newton method, e.g. 85, 1096
  and the Levenberg-Marquardt
    modification, e.g. 661
Gauss-Poisson process, e.g. 233
GDP estimates and components 669
Geary-Khamis aggregation method
  669

Gebiet und Bevölkerung: Daten 584,
  1148, 1255
Geburtenrate 194, 268, 340, 966,
  1255, 1428
Geburten und Sterbefälle 161, 591,
  1079, 1171, 1255
Geburtsmonate: wie verteilen sich
  die Geburten über das Jahr?
  1167
Geburtstagsproblem 687, 1400
Geburtstagsstichproben 1167
Geburt, Tod, Krankheit, Emigration
  und Immigration als stocha-
  stischer Prozeß, e.g. 193, 874
Gehan's extension of the Wilcoxon
  test 144, 770, 834, 893
Geheime Nachrichtenübermittlung
  91, 92, 187, 606, 696
Geigy-Tabellen 1426
GENCAT, introduction to, e.g. 1001
Generalized:
  and weighted least squares 300,
    658, 1383
  and conditional inverse, e.g. 443
  inverse regression 555, 663, 843,
    956, 1350
  inverses and their use in the linear
    model, e.g. 54, 716, 1207, 1390
  Lagrangian function method
    applied to optimal systems
    reliability 1308
  least squares, e.g. 732
  linear models 6, 31, 288, 346, 409,
    520, 573, 587, 674, 732, 814,
    863, 864, 922, 936, 1122
  Linear Models, GLIM 288, 409,
    863, 864
  linear models: for covariance ad-
    justment and standardization
    751
Generalized linear models in
  epidemiological research: case
  control studies 6
Generalized:
  reduced gradient methods applied
    to optimal systems reliability
    1308

Wilcoxon:
  two-sample tests for the
    comparison of survival
    curves 770
  test for comparing two sets of
    arbitrarily right censored
    data 770, 834, 893
General linear model, e.g. 31, 1284
  see also linear models and
    generalized linear models
  and direct standardization: a
    comparison 805
  approach to the analysis of
    variance, e.g. 674
  Bayesian inference for the 143
General-purpose simulation system,
  GPSS, e.g. 760, 849, 1375
Generating:
  continuous and discrete random
    variables by the inverse-
    transform approach, e.g. 760
  functions, e.g. 454, 643, 658, 874,
    1028, 1315
  pseudo-random normal deviates
    421, 1035
  specific random variables, e.g. 661,
    760, 1085, 1117, 1143
  uniform and non-uniform random
    variables, e.g. 661, 908
Generation of variates:
  from distribution tails 1186
  from 13 specific distributions 661
Generations, cohorts and conformity
  851
Genetic and cancer hazards 84
Genetics:
  mathematical theory of quantita-
    tive 159
  statistical methods in 331, 538, 852,
    1093, 1095
Genetik:
  kontinuierlich variierender Merk-
    male 852
  Wahrscheinlichkeitsmodelle und
    statistische Methoden 331,
    538, 852, 1093, 1095
Genotype distributions for relatives

in randomly mating populations
  331
Genotypes and phenotypes in experi-
  mental populations 331
Genotyp-Umwelt-Wechselwirkun-
  gen 852
Gen-Wechselwirkungen 852
Geoffrion implicit enumeration 1308
Geographie, statistische Methoden in
  der 389, 968, 1118, 1333, 1425
Geometric programming, e.g. 860,
  1308
Geometry: computational geometry
  and statistical applications 97
Geordnete Kategorien in Kontin-
  genztafeln 426, 427, 428, 861,
  1027
Geostatistics: sampling of particulate
  materials 469
Geowissenschaften: statistische
  Methoden 389, 469, 968, 1118,
  1333
Gerichtete Daten (z.B. Vogelflug)
  und ihre Analyse 57, 364, 839,
  840, 1029, 1373
German language, some hints 110
Gerontology: see elderly people
Geschichte der Statistik 351, 473, 880,
  1006, 1037, 1039, 1218, 1317
Geschichtete Stichproben 21, 98, 302,
  621, 729, 1061, 1137, 1201, 1274,
  1433
Gesetze der großen Zahlen 643, 768
"Gesetz seltener Ereignisse": Verall-
  gemeinerung 1394
Gesundheit der Bevölkerung der
  Bundesrepublik: Daten 161,
  163, 1079, 1255
Gesundheitswesen der Bundesrepu-
  blik: Daten 161, 163, 1079, 1255
Gewichtete Regression nach Grizzle,
  Starmer und Koch: siehe GSK-
  Ansatz
Gewogene kleinste Quadrate: siehe
  GSK-Ansatz
GFREG, generalized F distribution
  using maximum likelihood
  regression program 560

Ghosh-Ansatz zur zweidimensionalen Gitterschichtung bei proportionaler Aufteilung 302
Givens transformations, e. g. 661, 1205
GK aggregation method 669
Gleichverteilte Zufallszahlen, z. B. 45
Gleitende Durchschnitte, z. B. 942
GLIM 288, 409, 863, 864
    3 and PRISM 409
Goal scoring and the negative binomial distribution 1063
Gödel's theorem, e. g. 1113
Gompertz:
    distribution, e. g. 332
    model, e. g. 1098
        data plotting 566
Goodman and Kruskal's:
    gamma, e. g. 429, 836, 1110
    lambda, e. g. 429, 1110
    measures of association: gamma and tau, e. g. 429, 1110, 1334
    tau, e. g. 429, 1110
Goodman's gamma, e. g. 429, 836
Goodness of fit 41, 115, 643, 658, 760, 770, 793, 821, 836, 1049, 1072, 1153, 1290
Goodness-of-fit:
    binomial distribution, e. g. 1290
    loglinar models, e. g. 115, 1110
    lognormal distribution 647, 735, 760, 761, 770, 952
    normal distribution 290, 658, 806, 821, 836, 1153
    Weibull distribution, e. g. 41, 735, 952
Goodness of link tests 34
Government:
    statistical service (U. K.) 913
    statistics of the United Kingdom 1106
GPSS: Simulation with GPSS, e. g. 424, 760, 849, 1033, 1375
Graduate programs in statistics, e. g. 1149
Graeco-Latin square, e. g. 121, 674
Gram-Charlier function to fit reference distributions, e. g. 441

Gram-Schmidt:
    method, modified, e. g. 661
    orthogonalization, e. g. 54, 1205
Graphen, gerichtete: z. B. Petri-Netze, siehe dort
Graphical:
    displays 3, 23, 47, 56, 90, 168, 183, 200, 202, 342, 351, 363, 417, 520, 659, 683, 728, 755, 828, 866, 915, 916, 952, 1114, 1188, 1189, 1209, 1239, 1317, 1319, 1320, 1366, 1373, 1444
        of multivariate data 47, 69, 70, 97, 342, 351, 370, 683, 1209, 1317, 1359, 1366
        of means 23, 200
    methods:
        see also graphische Methoden
        exploring distributional models for data, e. g. 417
        for comparing two or more sets of numbers, e. g. 417, 824
        in nonparametric statistics 363
        of diagnostic regression analysis 34
        visual display of multivariate data 47, 342, 351, 370, 659, 1317, 1359, 1366
    perception 200, 728
    test for multivariate normality 342, 634
    two-way table diagnosis 127
Graphics:
    and interactive graphics, e. g. 829
    software, e. g. 168, 520
Graphik-Software, z. B. 168, 520
Graphische:
    Darstellungen 3, 23, 47, 56, 90, 200, 342, 683, 728, 755, 1114, 1188, 1189, 1209, 1317, 1319, 1320, 1366, 1444
    Darstellung: Manipulationen 3
    Methoden 47, 56, 69, 168, 183, 200, 202, 342, 351, 363, 417, 520, 659, 683, 755, 828, 866, 915, 916, 952, 1209, 1239, 1317, 1319, 1320, 1359, 1366, 1373

Graphs:
  games with, e. g. 1315
  purposes of 90, 351, 728, 1317, 1319, 1320
Graph-theory, e. g. 1315
Graeco-Latin squares, e. g. 121, 1408
Greenwood's formula and actuarial estimator, e. g. 234
Grenzwerte von Folgen und Funktionen, z. B. 56
Grizzle-Starmer-Koch, Modell nach (GSK-Ansatz) 374, 519, 681, 798, 1001, 1110, 1122, 1279
Group-divisible designs, e. g. 1036
Grouped and matched data: classical methods of analysis, e. g. 137
Group divisible design, e. g. 1036
Grouping variables according to their interrelationship: see factor analysis
Group:
  mixtures screening, e. g. 160
  sequential designs 694, 1058, 1182, 1393
  testing:
    extensions of the Dorfman procedure 592, 812, 923, 1051
    for modified binomial model, Dorfman-type 1051
    for binomial estimation 592, 812, 923, 1051
    robust 592
    two- and multistage 923
    with/without retest 923
Growth:
  and deterioration of reliability (repairable systems) 32
  and migration in human population, e. g. 1281
  and response curves, computer program for the distribution-free analysis of 703
  curve analysis 178, 422, 566, 627, 666, 690, 703, 716, 1065, 1209
  curves:
    and response curves 21, 44, 178, 300, 422, 566, 627, 666, 690, 703, 1065, 1209
  classification of 717
  random coefficient regression models, two-stage estimation procedures 627
  functions, e. g. 85
  models, nonlinear 178, 300, 566, 666, 703, 1065, 1209
  plant 178, 666
  trajectories, mean, classes of 21
Gruppenbildung durch Cluster-Analyse, siehe dort
Gruppensequentielle Pläne 694, 1058, 1393
GSA analysis flexibility: e. g. a test for marginal symmetry 1110
GSK:
  analysis 374, 681, 798, 1001, 1110, 1279
  -Ansatz oder GSK-Modell 374, 519, 681, 798, 1001, 1110, 1122, 1279
  approach 374, 681, 798, 1001, 1110, 1279
  technique used to obtain estimates of the standard errors of measures of association and partial association 724
Guttman's lambda measures, e. g. 1334

Hadamard matrix designs 270
Hämatologische Studien, z. B. 1353
Half-normal plot 249, 300, 1440
Hampel M-estimators 547, 1248
Haploidie und Polyploidie, z. B. 852
Happenstance data, problems in the analysis of 121
Hardy-Weinberg law 331, 643, 836
Harmonic:
  analysis, Fisher's test on the significance in, a Monte Carlo computer program for the critical values 1144
  curves, plotting of 342, 351, 840

Harmonisches Mittel, z. B. 961
Hartley-Sielken model in low-dose extrapolation, e. g. 198
Hart's:
    difference score, e. g. 947
    feeling-of-knowing prediction, e. g. 947
Hat matrix 219, 442, 547, 548, 1264
Hauptkomponenten-Analyse 8, 54, 83, 370, 501, 513, 604, 634, 661, 767, 800, 814, 831, 835, 840, 855, 858, 922, 1078, 1085, 1284, 1436
Haushaltsmitglieder-Befragung, z. B. 1169
Haushalts- und Familienstruktur: Daten 161, 591, 1079, 1255
Hazard:
    function: see hazard rate functions
    plotting 332, 437, 770, 950, 952
    rate function and expected duration of strikes 763
Hazard rate functions 99, 173, 234, 265, 332, 437, 761, 763, 770, 950, 952, 1065, 1348
Hazards:
    model, multiplicative log-linear, e. g. 234
    proportional 11, 144, 234, 649, 651, 770, 862, 1060
    step function, proportional hazards, e. g. 437
Hb values: modeling the distribution of, in a population of nonpregnant women containing normal nonanemics contaminated with anemics in the lower tail 149
Health data reporting, summary rates for, e. g. 194, 380
Healthy population: reference values in laboratory medicine 441
Heirats-, Ehe- und Geburtenstatistik, z. B. 340
Heiratstafeln: Generationentafelkonzept und Periodentafelkonzept, z. B. 340
Heterogeneity, problems of . . . in demography 750

Heteroscedasticity in regression analysis 955, 956, 1350
Hewlett-Packard HP-41C: Algorithmen der Statistik 679
Hierarchical:
    classifications, e. g. 608
        see also cluster analysis
    clustering versus stepwise regression 442
    data, a distribution-free test for symmetry in 94
    designs, hierarchal designs, nested factors 123, 674, 896, 1239, 1408
Hierarchische Modelle 123, 674, 896, 1239, 1408
Highway:
    risks of the 1204
    maintenance, application of Markovian decision theory to the problem of 999
Hinge and trimean, e. g. 1319, 1320
Histocompatibility testing, e. g. 331
Histogramm, Alternativen zum klassischen 183, 401, 490, 1042, 1206, 1319, 1320, 1339, 1345, 1359
Histograms, alternatives to 183, 401, 490, 1042, 1206, 1319, 1320, 1339, 1345, 1359
Histogramme: Beispiele 1040, 1042
Histograms:
    examples 1040, 1042
    on optimal and data-based 1206
    smoothing . . . by means of lattice and continuous distributions 401
Historische:
    Demographie, z. B. 1404
    Kontrollen, z. B. 1313
History:
    of probability and statistics: see history of statistics
    of statistics 351, 473, 880, 1006, 1037, 1039, 1218, 1317
        notes of Sheynin on his papers 1218
HLA, genetic studies of . . . and cancer, statistical methods for 225

Hochrechnungen, z. B. 727
Hochschulstatistik: Daten 161, 1255
HOD, honest ordering difference,
    with tables 111
Holm procedures 565, 1180, 1242
Holt-Winters methods with listed
    computer programs, e. g. 4
Homogeneity:
    chi-square test of homogeneity of
        distributions 103, 836, 1153,
        1334
    distribution-free tests of, with
        dependent samples 709
    marginal, tests for . . . in square
        contingency tables 710
    of distributions, chi-square test of
        103, 836, 1153, 1334
    of proportions, test of 103, 352,
        353, 355, 1152, 1153, 1334,
        1335
Homogenitätstests:
    bei abhängigen Stichproben, ver-
        teilungsfreie 709
    Chiquadrat-Homogenitätstests für
        k unabhängige Stichproben
        103, 352, 836, 1152, 1153, 1334
Homogenität:
    zwischen Untersuchungseinheiten
        oder Individuen, jeweils als
        Blöcke auffaßbar, die nicht
        gleich stark besetzt zu sein
        brauchen 94
    zwischen Verlaufskurven, die als
        unterschiedlich stark besetzte
        Blöcke auffaßbar sind, Test
        der 94
Homoscedasticity of residuals, e. g.
    1284
Honest ordering difference, with
    tables 111
Hot-deck procedures, e. g. 820
Hotelling's T^2, e. g. 634, 800, 840,
    1089
    the sample geometry of Student's t
        statistic and 634
Householder transformations, e. g.
    661, 1205

Household surveys, assessment of
    errors in developing countries
    1342, 1343
How long is the wait for an uncom-
    mon event? 1145
HPD-Intervalle (highest posterior
    density) 684
HSD technique, e. g. 111
H-Test:
    (Kruskal-Wallis Test) 285, 770, 793,
        836, 1152, 1153
    Approximationen und kritische
        Schranken 402, 598, 600
H test: extensions of 509, 578, 770
H-Test:
    kritische Schranken 402
    lineare Kontraste 787
    Verallgemeinerungen 509, 578, 770
Huber M-estimators, e. g. 547, 1248
Human:
    blood groups, e. g. 331
    communication research, multi-
        variate techniques in 900
    population:
        growth and structure of, in the
            presence of migration 1281
        see also Demographie and Mi-
            gration, and Bevölkerungs-
            statistik
Hunter's inequality 59
Hypergeometric:
    and binomial sampling plans 984
    distribution, e. g. 542
        generalizations 632
    functions of matrix arguments, e. g.
        1078
Hypothesis testing, e. g. 48, 66, 229,
    231, 572, 643, 655, 658, 769, 892,
    1298, 1413
    overall and multiple, e. g. 59
    with complex distributions 892

**IDA, Interactive Data Analysis, con-
    versational statistics with** 47, 69,
    70, 1121

Ideas, sources of, e.g. 123, 1318, 1320, 1322
Identification:
  and classification: see cluster analysis
  problem: a supply-demand model, e.g. 1420
Identifying exceptional data points: the hat matrix 219, 442, 547, 1264
Image:
  analysis, e.g. 377
    and recognition, structural methods in 717
    and stereology 1118
  models 717
  texture survey 717
IMA processes, e.g. 122
Immigration and migration, e.g. 193, 1281
Implementation strategies for statistical methods, e.g. 520
Implicit functions, e.g. 85
Importance functions, e.g. 524
Important variables, selection of 858
Impurities in distributions, detection of 343
Imputation in surveys 248, 714, 820, 1125, 1161
Inbreeding and nonrandom mating, e.g. 331
Incidence:
  data, cyclic trends in, a significance test for 1126
  density data in cohort studies, analysis of 136
  rates 396, 682, 1182
    and ratios, use of Poisson regression models in estimating 386
    comparison of two, e.g. 146
    goodness of fit, e.g. 1124
Inclusion-exclusion, e.g. 1315
Income:
  and wealth distribution 1106
  distribution models 30
Incomplete beta function:
  and residuals e.g. 235
  table e.g. 406

Incomplete:
  block designs 251, 1257
  data in sample surveys 248, 714, 820, 1161
Incubation period of disease, median 27
Independence:
  of random variables, e.g. 454
  tests of 10, 658, 793, 836
Independent tests, combination of 414, 588, 719
Index numbers 20, 43, 349, 614, 669, 1106, 1171, 1224
  in international comparisons, e.g. 669
  log-change 614
Indexzahlen 20, 43, 268, 349, 614, 669, 1106, 1171, 1224
Indicator (dummy) variables in regression analysis 4, 300, 564, 798, 902, 955, 1383, 1437
Indices for diagnostic screening 79
Indirectly adjusted rates 103, 194, 380
Induction:
  Induction, Induktion 182, 291, 473, 684, 737, 882, 1401
  popular . . . and improving inductive inference 640
Induktive Statistik, z.B. 121, 231, 291, 304, 473, 817, 1152, 1153, 1238
Industrial experiments:
  robust regression in, e.g. 765
  the design of effective 270
Industrial firm size distributions, e.g. 50
Industrie-Statistik, Statistik des produzierenden Gewerbes 20
Inequalities 454, 1028
Inequality and Lorenz curves 30
Infant mortality, e.g. 166, 194
Infectious disease statistics, mathematical models for 35
Inference:
  robust, the Fisherian approach 756
  statistical 121, 229, 231, 304, 817, 1153, 1209, 1238, 1322

threats to valid inference-making, e.g. 220

Inferential statistics, e.g. 48, 121, 286, 1153, 1238

Infinitely divisible point processes, e.g. 233

Influence curve, e.g. 488, 547, 580, 1097

  sample and empirical, e.g. 219

Influential data, influence curves for estimators 75, 219, 547, 1264, 1316, 1340, 1358, 1424

Influence functions and regression diagnostics 75, 219, 755, 990, 1264, 1340, 1383

Influential observations:

  in regression analysis 75, 219, 300, 442, 547, 755, 956, 990, 1264, 1316, 1340, 1383

  in the analysis of variance 1043

Influential subsets in regression 442

Information:

  and Dokumentation 399, 746

  flows: Petri nets 410, 529, 1016, 1104, 1105

  processing by Petri nets 410, 529, 1016, 1104, 1105

  Retrieval 399, 746

Informationsfluß-Beschreibung anhand von Petri-Netzen 410, 529, 1016, 1104, 1105

Informationsverarbeitung in der Medizin, z. B. 692

Information:

  transmission and information transformation: Petri nets 410, 529, 1016, 1104, 1105

  und Dokumentation 399, 746

Inheritance models, e.g. 331

Input-output:

  analysis 894

  matrices, optimal triangulation of large real world 456

  process: see queueing models and epidemic models

  -Rechnung 20, 268, 456

Inside-out plot 1359

Inspection:

  areas and inspection times, e.g. 781

  policies, analytical considerations on 999

Instantaneous failure rate: see hazard rate function

Instrumental drift, e.g. 733

Insurance statistics 64, 431, 528, 575, 758, 1300

Integer Programming, e.g. 536, 1308

Integrated autoregressive moving average models = autoregressive integrated moving average models, ARIMA models 185, 186, 255, 859, 1184

Integrating:

  research findings across studies: see meta-analysis

  studies, errors and limitations in 414

Integration of studies and significance levels: see meta-analysis

Intelligent statistical software and expert systems, more, e.g. 478

Interaction:

  between two and among three or more species, e.g. 1065

  in experimental design 131, 232, 417, 525, 674, 716, 789, 1036, 1122, 1390

  in two-way classification models, inference on the structure of, e.g. 716

  row plot and column plot in two-way tables 127

Interactions:

  identifying the dominant partner of two 1365

  in contingency table analysis, e.g. 1311

  predator-prey and host-parasite, e.g. 1065

  two-way and three-way, decomposition of, in analysis of variance 417

Interactive:

  data analysis 47, 69, 70, 1121

Interactive:
  graphics for model assessment,
    e.g. 47
Intercept:
  parameter testing in one-sample
    two-sample and k-sample
    linear regressions by
    adapting simple non-
    parametric methods 819
  rank tests for ... in linear regres-
    sion problems 819
Interclass correlation, a nonparametric measure of 1221
Interdependence and covariances
  1323
Interface, symposium on the: computer science and statistics 520
Interior analyses in fitting equations
  to data 249
Interlaboratory:
  studies: robust statistical analysis
    of 1123
  testing 733, 832
International comparisons of real
  gross domestic products 669
Internationale Übersichten: Daten
  1255
Interpersonal skill in statistical consulting 477, 888, 1149, 1257, 1441
Interpretation:
  and presentation of statistical
    results in medicine 888
  of data and results, e.g. 229, 328,
    379, 888, 916, 1151–1153,
    1263, 1318–1323
Interrupted time series analysis 500,
  865
Interval estimation, e.g. 66, 658, 769
Intervallschätzung, z.B. 66, 658, 769
  des größten Mittelwertes von k
    Normalverteilungen, z.B. 189
Intervals
  from normal likelihoods, e.g. 643
  from significance tests, e.g. 643
Intervention analysis, e.g. 4, 865
Interventionseffekt-Analyse bei Zeit-
  reihen-Experimenten, z.B. 4,
    865

INTERVERSA-Studie 967
Interview:
  conducting an 36, 125, 278
  (persönliches) und Telefon-Erhe-
    bung 384, 459, 497, 1200,
    1276, 1325
  or survey by telephone: see tele-
    phone interview
Interviews, personal interviews versus survey by telephone 384,
    459, 497, 1200, 1276, 1325
Intra- and interlaboratory variability,
  e.g. 605, 1123
Intraclass correlation 745, 798, 1138,
  1220, 1221, 1233, 1238, 1253,
  1380, 1408
  a nonparametric measure of 1138
  effects on covariance analysis 1233
  estimation of, for the analysis of
    covariance model 1253
  nonparametric measures of 1138,
    1220
  selecting the population with the
    largest, e.g. 406
  simulated 1380
Intraclass dependence, tests of ... for
  qualitative variables 60
Intra-individual and inter-individual
  variation, e.g. 441
Introduction to statistical analysis
  286, 1153, 1238
Inventory models, e.g. 760
Inventur auf Stichprobenbasis 302,
  1172, 1179
Inverse:
  Chi-Verteilung 684
  Gaussian distribution 39, 184, 372,
    763, 847, 1028
  tables, standardized, of the
    cumulative probability
    function 184
Inverse Gaussian distribution,
  the three-parameter: zero mean,
    unit variance, third standard
    moment as shape parameter,
    tables of the cumulative
    probability function 184

median estimation 1324

of the normal cumulative distribution function, alternatives to Hastings' approximation to the 39

regression: see calibration

Inverses, generalized, and their use in the linear model, e.g. 54, 1207

Inverse-transform approach for generating random variables, e.g. 760

Investment choice 398, 654, 1119

Inzidenzraten 682, 888

Vergleich zweier, z. B. 146

Irwin-Fisher-Test: siehe Fisher-Test, exakter

Isodynes for $\bar{X}$ and R control charts 781

Isolining problems, e.g. 1333

Isopleths, e.g. 1333

and modes, estimating both 1155

ISPAHAN (a statistical package), e.g. 1052

Iterative:

proportional fitting, e.g. 472

weighted least squares algorithm, e.g. 936

Jacknife techniques 160, 325, 326, 915, 916, 1023, 1419

James-Stein:

and ridge estimators, e.g. 31

estimator, Stein estimator, e.g. 175, 327

Jephtha's sacrifice, a game theoretic approach 128

JKW model, after Jöreskog, Keesling and Wiley 814

Job-shop model, e.g. 760

John's three-quarter fractional factorials 270

Johnson system of distributions, e.g. 541

Joint dependence, e.g. 1323

Jonckheere-Test 636, 766, 976, 977, 1069, 1227, 1294

Jordan canonical form, e.g. 749

Judge, convincing evidence ("beyond a reasonable doubt") for a 299, 354, 356, 959, 1241

Judgement under uncertainty: bias and bias correction, e.g. 640

Justice:

distribution, expected 613

the sense of, as a unifying principle of social life: distribution of justice sentiments experienced by the members of a social aggregate 613

K^2-test and W-test, e.g. 1052

Kalibrierung 148, 300, 318, 319, 377, 395, 577, 589, 610, 685, 733, 815, 990, 1176, 1280, 1295, 1324

see also calibration

Kalman filter, e.g. 4, 769, 811

from the perspective of Goldberger-Theil estimators 271

Kanonische Korrelation 8, 93, 114, 411, 482, 501, 661, 686, 731, 767, 798, 800, 835, 840, 855, 922, 1078, 1284, 1303, 1324, 1350, 1381

Kaplan and Meier's product-limit method program: survivorship function, percentiles and survival curve plotting in linear and semilogarithmic scales 770

Kaplan-Meier's estimator, e.g. 144, 194, 234, 770

Kappa coefficient and extensions, e.g. 367, 706, 766, 1110

Kategoriale Regression, z. B. 346

"Kategorische" Daten und ihre Analyse 103, 334–336, 352, 353, 355, 374, 425, 490, 519, 658, 663, 766, 798, 814, 822, 833, 965, 1056, 1170, 1425

Kaufkraftparität, z. B. 349

Kaufverhalten, Auswahl beim, z. B. 347

Kausale Analyse anhand von loglinearen und Logit-Modellen, z. B. 352

Kausalität 21, 662, 882, 1139
siehe auch causal models

K-Clustering as a detection tool for influential subsets in regression 442

Kendall's:
coefficient of concordance, e. g. 285, 836, 1110
rank correlation 285, 440, 766, 793, 836, 1110
S, nomogram for obtaining z-test statistic from ... and sample size 10 to 50   440
tau a and Kendall's tau b as well as tau c, e. g. 1110
tau and gamma and mean ridit, e. g. 611

Kernel estimators 490, 1071

Kinderlosigkeit in der Ehe nach dem Mikrozensus von 1981 519

Kindersterblichkeit, z. B. 166

Kinetic tree theory, program, e. g. 524

Klassierung von Daten, statistische Klasseneinteilung, z. B. 1070

Klassifikation, automatische: see cluster analysis

Kleinrechner in der Statistik, Programmpakete für 404, 679, 741, 918, 933, 1423

Kleinste Quadrate, Methode der kleinsten Quadrate 249, 300, 468, 490, 658, 733, 769, 814, 890, 1390

Klinisch-chemische Daten:
Abweichungen vom Typ einer Normalverteilung für 20 Serumbestandteile 1427
multivariate Datenbeurteilung 1103

Klinische Daten: Norm und Normabweichung 390

Klinische Studien:
(Therapiestudien) 888, 1348
Planung von Erhebungsbogen und Datenbank 776

Zwischenauswertungen 692, 694

Klumpen-Stichproben 98, 205, 621, 729, 779, 1061, 1137, 1274, 1431, 1433

Knapsack problem 1308

Knobeleien und Tüfteleien für numerisch Interessierte 705

Knock-out tournament 406

Knowledge:
Base Management System 1108, 1163
representation and expert systems: see expert systems

Kodiervariablen und Kodiertechnik, z. B. 1122

Kohorten-Analyse 47, 138, 161, 194, 225, 282, 332, 338, 367, 416, 422, 682, 791, 851, 1182, 1230, 1292, 1404, 1428

Kohortenansatz nach Whelpton, z. B. 340

Kohortensterbetafel für Deutschland, ansatzweise Berechnung der 282

Kohorten-Studie: siehe Kohorten-Analyse

Kollinearität 75, 551, 831, 1316, 1340, 1383

Kolmogoroff-Smirnoff-Anpassungstest:
diskrete und gruppierte Daten 1049
graphischer Test 821

Kolmogoroff-Smirnoff-
1-Stichprobentest: kritische Schranken 648
2-Stichprobentest 214, 793, 836, 1072
k-Stichprobentest: kritische Schranken 1417
Tests: siehe auch Kolmogorov-Smirnov tests

Kolmogorov-Smirnov tests 214, 648, 760, 793, 821, 836, 1049, 1072, 1417

Kombination unabhängiger χ^2-Tests 704

Kombinationseffekte in der Arznei-
mittelforschung 1353, 1332
Kombinatorik 484, 617, 705, 1315
Kommunikationsabläufe in der
Medizin und ihre Darstellung
durch Petri-Netze 86, 692
Komplexe Abläufe, ihre
Modellierung durch Petri-
Netze: siehe dort
Konfigurationsfrequenzanalyse
1122, 1347
Konfirmatorische Datenanalyse
1318, 1322, 1323, 1347
Konflikt-Modelle 637, 1092
Kontingenztafel: see also
contingency table
Kontingenztafel-Analyse 24, 93, 103,
114, 115, 123, 214, 263, 285, 288,
334-336, 352, 353, 355, 374,
425-429, 519, 653, 658, 660, 653,
793, 798, 836, 854, 855, 861,
1012, 1026, 1056, 1100, 1101,
1110, 1334, 1335, 1357, 1384,
1425
siehe auch Korrespondenzanalyse
bei geordneten Kategorien 10, 204,
263, 1027, 1357
Kontingenztafeln: Analyse:
der Residuen 472, 620
dreidimensionaler Tafeln 103, 334,
335, 425-429, 836, 1056, 1101
Kontingenztafeln:
duale Skalierung 965
Fisher-Test für $r \times c$ 877, 1012
odds ratio in 1110
quadratische: 710, 861, 1449
Prüfung der Homogenität der
Randverteilungen 710
simultane Vergleiche mit einer
Kontrolle 1026, 1141
Wechselwirkungen in, z. B. 1311
Kontinuitätskorrektur, z. B. 470
Kontrollen: wie viele? 388
Kontrollkarten 924, 1329, 1389
Konzentrationsmessung und Lorenz-
kurve 20, 43, 268, 349, 1171
Korrelation 8, 210, 739, 798, 1316

see also correlation
multiple 8, 210, 771, 798
Korrelationsanalyse 8, 210, 739, 798,
902, 1316
gemeinsame Betrachtung vieler
Korrelationen, z. B. 1316
Gleichheit abhängiger Korrela-
tionskoeffizienten 77, 934,
1256, 1418
Korrelationskoeffizient:
Produktmoment-
Korrelationskoeffizient
approximative Vertrauensbereiche
116
kritische Werte 981, 987, 988
Robustheit des t-Tests gegenüber
Nichtnormalität 321
Transformationen, z. B. 1273
Korrelationskoeffizienten:
Gleichheit abhängiger 77, 934,
1256, 1418
Gleichheit zweier, z. B. 1297
Kombination unabhängiger, z. B. 1344
Korrelationsmaße und Assoziations-
maße 81, 349, 429, 739, 1154
Korrespondenzanalyse 47, 164, 448,
535, 596, 608, 612, 767, 800, 809, 965
als Hauptkomponentenanalyse
von Profilen 596
als Spezialfall der kanonischen
Korrelationsanalyse 596
Kosten-Nutzen-, Kosten-Wirksam-
keits- und Kosten-Nutzwert-
Analyse, z. B. 1033
Kostenoptimale Stichproben-
Prüfpläne beim Ziehen ohne
Zurücklegen, z. B. 211
Kovarianzanalyse 21, 93, 100, 210,
251, 272, 285, 286, 288, 412, 573,
585, 674, 732, 751, 780, 857, 955,
1073, 1093, 1095, 1121, 1122,
1122, 1142, 1209, 1219, 1233,
1238, 1253, 1284, 1390, 1398,
1408
siehe auch ANCOVA
Rang- und robuste Methoden 100,
412, 585, 764

Kovarianzmatrix in linearen und nichtlinearen Modellen, z. B. 587

Kovarianzmatrizen, der Vergleich von 31, 370, 587, 712, 814

Kovarianzauswahl 1347, 1384

Kovarianzselektion:
als explorative Methode 1347, 1384
und log-lineare Modelle 1384

Kovarianzstruktur, z. B. 513, 587

Krebsforschung: statistische Methoden 173, 612, 888, 1021, 1060

Krebsrisiken 84, 292, 338, 418, 713, 917, 1021

Krebssterblichkeit in der Bundesrepublik 162

Kreis-Diagramme und Weiterentwicklungen 57, 839, 1373

Kreis und Kugel, Statistik für 57, 364, 839, 840, 1029, 1373

Kreis: Ungleichungen vom Tschebyscheff-Typ auf dem Kreis 839

Kruskal-Wallis one-way analysis of variance 285, 770, 793, 836, 1152, 1153
extensions of 509, 578, 770

Kruskal-Wallis statistic 285, 770, 787, 793, 836, 1152, 1153
approximations and critical values 402, 598, 600

Kryptographie 91, 92, 187, 606, 696

K-sample rank test for complete block designs with proportional cell frequencies 509

k-Stichproben-Tests für geordnete Alternativen, z. B. 82

K^2-test and W-test, e. g. 1052

Kuder-Richardson formulas, e. g. 739

Künstliche Intelligenz und Expertensysteme, z. B. 1163

Kuhn-Tucker conditions 1308

Kurtosis 630, 739, 1040, 1152, 1153, 1390
and skewness, measures of multivariate, e. g. 840

"Kurzgeschichten", auch für Statistiker 568, 640, 705

L_1-norm regression, e. g. 273, 929

Laboratories: measurement agreement comparisons among standardizing laboratories, e. g. 733

Laboratory medicine, reference values in 441

Labordaten-Variabilität, z. B. 605

Lagerhaltung, z. B. 347, 1052

Lagrange multipliers and Kuhn-Tucker conditions in optimal systems reliability 1308

Lambda:
measure of association 81, 103, 349, 429, 766, 1056, 1110
Wilks', e. g. 501, 1284

Lancaster's definition of interactions in three-dimensional contingency tables, e. g. 1311

Landesstatistikämter und das Statistische Bundesamt 554

Landry-Konzept für das Bevölkerungswachstum, z. B. 340

Land- und Forstwirtschaft, Fischerei: Daten 1255

Langfristprognostik, z. B. 150

Largest mean:
in repeated measurements design, confidence interval for the 190
of k normal populations, interval estimation of the 189

Larynx-cancer survival, e. g. 649

Laspeyres-Paasche index numbers, e. g. 614

Lateinische Quadrate 251, 573, 674, 955, 1408

Latent:
-class model, e. g. 344, 472
root regression, e. g. 300
roots, e. g. 54
structure analysis 17, 425, 603, 814, 1078, 1347

exploratory, e. g. 425
variable models for categorical
data 344
variables:
see also LISREL models
multivariate analysis with 717
Latin:
square design 121, 251, 573, 674,
955, 1001, 1408
squares, Graeco- and Hyper-
Graeco, e. g. 121
Lattice:
data, e. g. 1118
designs, e. g. 538
sampling, e. g. 621
Law:
and statistics 299, 354, 356, 1241
of small numbers, belief in the 640
of the iterated logarithm, e. g. 454
Laws of large numbers, e. g. 454, 643,
768, 1028
Learning theory: a two-parameter
likelihoods example 643
Least absolute:
error (LAE) regression, e. g. 273,
929
residual (LAR) regression 273, 547,
929
value:
estimation in regression models:
an annotated bibliography
275
(LAV) regression 273, 929
regression and linear
programming 275
Least:
favorable configuration or best
candidate selection procedures, e. g. 406
squares:
alternating, e. g. 1436
estimation 249, 300, 468, 490,
658, 733, 769, 814, 890, 955,
956, 1390, 1420
constrained, e. g. 443
of regression coefficients for a
subset of predictors and its
bias 890

scaling, e. g. 513
two-stage, e. g. 1420
weighted 300, 798, 936, 1383
Lebensdauerstudien 173, 184, 234,
242, 332, 625, 649, 761, 770, 864,
893, 952, 974, 1060, 1168, 1198,
1226
Lebensmittelprüfung, sensorische
618
Lebensversicherung: Risiko-
Klassifizierung, z. B. 243
Ledermann's life table coefficients,
e. g. 1428
Legal probability 299, 354, 356, 959
Length-of-stay in hospital and
duration of strike 763
Lengths of series until at least k
events have taken place, given
negative binomial distribution
with the probability p of this
event 1145
Lepage-Test 782
Lernversuche mit Ratten, z. B. 1353
Leslie matrices in demography, e. g.
54
L-estimators, e. g. 547, 719
Letter-value displays 547, 1339
Letter values or quantile values for
data summary, e. g. 547, 1025
Level and shape comparisons of two
response curves 703, 708
Levenberg-Marquardt method for
the nonlinear regression
problem, e. g. 661
Leverage in regression:
models 75, 174, 219, 442, 547, 550,
1316, 1340
through the origin 174
Levinson-Algorithmus 906
Lexikon: Wahrscheinlichkeitsrech-
nung und Mathematische
Statistik 920
Lexis-Diagramm 340, 1404, 1428
"Liars and statisticians" 568
Life:
data analysis: see survival data
analysis

Life:
  expectancy:
    and survival rates, e.g. 194
    at birth in Germany 282
  insurance 64, 243, 528, 575, 758,
    1300
  -table analysis program with plots
    of the estimated survivorship
    function, hazard function,
    and density function 770
  table, antenatal, e.g. 194
  table construction:
    and analysis 166, 194, 228, 285,
      332, 340, 667, 750, 761,
      1065, 1226, 1232, 1378, 1428
    and analysis: see also survival
    a time series approach to 1378
  table:
    death rate and its standard error
      194
    for Drosophila melanogaster,
      e.g. 194
    for survival and stages of
      disease, e.g. 194
    functions, statistical inference
      regarding 194
  tables:
    abridged and complete, e.g. 194
    FORTRAN computer programs
      for the construction of
      abridged and complete life
      tables 194
    single and multiple decrement
      life tables, e.g. 332
  table: theory of, complete,
      abridged, and new life table
      as well as statistical inference
      regarding life table functions
      194
  test data, attribute, estimation
      methods based on, e.g. 847
  testing:
    and reliability data in
      engineering 952, 1309
    properties of some time-sequen-
      tial statistics in, e.g. 265
Lifetimes: distribution of 234, 372,
    761, 762, 952

Light-Margolin measure of
    association on the concept of
    total variation for categorical
    data 83
Likelihood:
  analysis for the logistic model, e.g.
    643
  factoring (of) the likelihood for
    missing data 804
  function and procedures closely
    related to it 643, 658
  inference, fiducial inference, and
    structural inference, e.g. 48
  methods 643, 658
  ratio statistic, e.g. 103
Likelihoods: conditional and uncon-
    ditional, e.g. 813
Likelihood, shape, and adaptive
    inference 123
Likelihoods, normal, interval from,
    e.g. 643
Likelihood und Bayes-Statistik, z.B.
    684
Limit theorems, e.g. 1028
  rank statistics and 719
Line:
  estimation, robust, with errors in
    both variables 147, 300, 547,
    580, 587, 641
  straight, fitting a: see regression
    analysis and regression line
Linear:
  algebra, numerical, e.g. 1085
  calibration: see calibration
  combination:
    selecting the best normal popu-
      lation based on a 406
    of observations, mean and
      variance of a 121
  contrast of means 160, 674, 787,
    1153, 1307
  curve fitting with both variables in
    error 147, 300, 547, 580, 587,
    641
  discriminant function: see
    Diskriminanzanalyse
  dynamic systems: Bayesian
    methods 143

equality constraints in regression
and loglinear models 1115
equations:
    and generalized inverses, e.g.
        443, 445, 1207
    solving, e.g. 54, 443, 1085, 1207
forms, e.g. 840
Linearizable curves, e.g. 249
Linear:
    least squares, the geometry of 300
    matrix equations, e.g. 749
    models 31, 93, 288, 374, 412, 443,
        468, 564, 573, 587, 674, 732,
        769, 864, 911, 922, 937, 955,
        1122, 1368
        see also generalized linear
            models, and multivariate
            linear models
        and non-orthogonal data 937
        Bayesian 143
        for the analysis of longitudinal
            studies, e.g. 1368
        nonparametric methods 719
        proportional 241
        structural change in, Bayesian
            analysis of 143
    probability model, e.g. 13
    programming, e.g. 333, 490, 536,
        860, 1158, 1308
        and least absolute value regres-
            sion 275
    regression:
        see also regression, linear
        model, review of statistical tests
            and procedures which aid
            the experimenter in
            determining lack of fit or
            functional misspecification
            associated with the
            deterministic portion of
            the proposed 935
    stationary and linear nonstationary
        models, e.g. 122
    transformations and matrices, e.g.
        54, 749
Lineare:
    Algebra, z. B. 1085

Gleichungen, z. B. 1085
Kalibrierfunktionen: siehe calibra-
    tion
Modelle 31, 288, 374, 412, 443, 468,
    564, 573, 587, 674, 732, 864,
    922, 937, 955, 1122, 1207
    multivariate, z. B. 31, 587, 1078,
        1207, 1209
    proportionale lineare Modelle,
        z. B. 241
    robuste Verfahren in linearen
        Modellen, z. B. 587
    und ihre Anwendungen im Op-
        erations Research, z. B. 310
    und ihre Verallgemeinerungen 6,
        31, 288, 346, 409, 443, 520,
        573, 587, 674, 732, 814, 863,
        864, 922, 936, 1122, 1137
    und nichtlineare Programmierung:
        siehe Operations Research
    Regression: siehe regression, linear
    Transport- und Umladeprobleme
        293
Linearizable functions with corre-
    sponding transformations, e.g.
    1153
Linearizing transformations, e.g. 83,
    300, 547, 916, 1001, 1153, 1320,
    1383
Linear models 31, 288, 374, 412, 443,
    468, 564, 573, 587, 674, 732, 864,
    922, 937, 955, 1122, 1207
    see also generalized linear models,
        and multivariate linear
        models
    use of generalized inverses in, e.g.
        54, 1207
Linear:
    panel analysis, e.g. 665
    probability modeling in contin-
        gency table analysis: see
        GSK-Ansatz
    programming, e.g. 490, 536, 1158,
        1308
    regression:
        see regression, linear
        see also regression lines

Linear:
  structural relations: see LISREL
  transformations and characteristic
      roots, e.g. 443
Lines, resistant 547, 1339
Link tests, goodness of 34
LINWOOD 249
LISREL model, e.g. 344, 628, 806,
      807, 848
Litchfield-Wilcoxon: ED_{50} determi-
      nation, e.g. 1286
L_1-norm regression, e.g. 273, 929
Location:
  and dispersion, measures of 658
  and scale:
      differences in two-sample
          comparisons: graphical
          methods in nonparametric
          statistics 363
      parameters, nonparametric tests
          for 782
  detecting changes in 59, 1175,
      1212, 1438
  estimates, robust 35, 547, 580
  estimators:
      see also robust estimators
      a comparison 547, 580, 916,
          1247, 1248
      robust and resistant measures/esti-
          mators 547, 580, 916, 1247,
          1248
Logarithmic:
  normal distribution: see lognormal
      distribution
  series distribution, e.g. 542
Logarithmische:
  inverse Chi-Verteilung, z.B. 684
  Normalverteilung 41, 207, 208,
      452, 475, 503, 504, 541, 631,
      647, 735, 760, 761, 763, 770,
      847, 898, 952, 1010, 1152,
      1153, 1215, 1280
Log-change index numbers 614
Log change, the only symmetric, ad-
      ditive and normed indicator of
      relative change 1310
Logistic:
  discrimination 717

  distribution, e.g. 504, 541
  and standardizing for
      comparison, e.g. 916
  assigning of centroids, e.g. 916
  model 11, 178, 300, 566, 666, 682,
      703, 1098, 1209, 1313
  a two-parameter likelihoods
      example with some ob-
      served survival data 643
  data plotting 566
  likelihood analysis 643, 813
  models:
      application to cross-classified
          categorical data, e.g. 352
      application to the analysis of or-
          dinal responses, e.g. 449
  regression 8, 18, 135, 137, 173, 219,
      227, 288, 682, 813, 1182, 1314
      diagnostics 1077
      in epidemiologic research, e.g.
          288, 682, 1182, 1314
      unconditional l.r. for large strata
          and conditional l.r. for
          matched sets 137
Logistic regression:
  see also Pregibon's diagnostics
Logistik:
  Rundreisen und Touren 293, 1052
  Standorte 294
Logistiksysteme 1052
Logistik: Transporte und Umlade-
      probleme 293, 1052
Logistische:
  Modelle 227, 300, 352, 367, 374,
      564, 682, 888, 1182
  Regression 8, 18, 137, 219, 227, 288,
      682, 813, 1182
  Verteilung, z.B. 504, 541
Logit analysis:
  see logit models
  following Goodman versus dum-
      my variable regression analy-
      sis, e.g. 425
  model following Goodman versus
      additive probability models,
      e.g. 425
Logitlineare Modelle 11, 13, 42, 752,
      862, 1056

Logit-linear models and standardized proportions, e. g. 805

Logit:
model for ordinal-ordinal tables and for ordinal-nominal tables, e. g. 10
models 11, 13, 21, 103, 352, 425, 472, 531, 564, 751, 799, 805, 864, 1056, 1122, 1182, 1279, 1331, 1360
regression see logitlineare Modelle and Logitmodelle

Logits:
cumulative, e. g. 11
standardized, e. g. 805

Logit versus loglinear models for analyzing ordinal categorical data, e. g. 10

Loglineare Modelle, Anpassungstests zur Bewertung und Auswahl 115

Loglinear hazards model, multiplicative, e. g. 234

Log-linear model, an algorithm for fitting a general 471, 472

Log-linear modeling, eight common errors in published applications of 1279

Loglinear models 11, 21, 42, 103, 115, 285, 288, 346, 352, 353, 355, 425, 427, 428, 472, 513, 562, 564, 798, 822, 862, 864, 888, 972, 1056, 1078, 1110, 1115, 1279, 1334, 1384
and latent-structure analysis 425
and standardized means, e. g. 805
calculating expected frequencies following Goodman, e. g. 1110
econometric and psychometric applications in categorical data analysis 355
estimation and partitioning, e. g. 103, 352, 353, 355, 425, 428, 1110
estimation by iteration, e. g. 103, 352, 353, 355, 425, 428, 1110
evaluation and selection of 115
for analyzing rates and survivorship 562
for frequencies and for odds, e. g. 425
for frequencies and for odds in which response variables and/or explanatory variables have ordered levels, e. g. 428
for frequencies and for odds: the analysis of dependence in cross-classifications having ordered categories 428
for multidimensional tables and model selection 323
for ordinal-ordinal tables and for multidimensional tables having at least one ordinal variable, e. g. 10
for three-way tables, e. g. 103, 352, 353, 355, 425, 428, 472, 1110
for two-way, three-way and higher-dimensional tables, e. g. 103, 352, 353, 355, 425, 428, 472, 1110
goodness-of-fit: normed and non-normed descriptive fit indices, cross-validation procedures, and predictive precision indices 115
homogeneous constraints for 1115
iterative proportional fitting for three variables, e. g. 103, 1110
linear equality restrictions in 1115
selection and evaluation of 115
software for, e. g. 114, 285, 288, 409, 428, 471, 472, 863, 864, 472

Log-logistic and linear logistic regression models in the analysis of breast cancer survival 437

Log-logistic distribution, e. g. 234

Lognormal data containing zeroes and left-censored values: estimation of the mean 1010

Lognormal distribution 41, 207, 208, 452, 475, 503, 504, 541, 631, 647, 735, 760, 761, 763, 770, 847, 898, 952, 1010, 1153, 1215, 1278

Lognormal distribution
  a bivariate semi-lognormal distri-
    bution, estimation in 1278
  analysis of coarsely grouped data
    from the 503
  confidence estimation of measures
    of location in the 898
  estimation in a bivariate semi-
    lognormal distribution 1278
  estimation of the mean for lognor-
    mal data containing zeroes
    and left-censored values 1010
  estimation of the three-parameter
    207, 208, 452, 647, 735
  interval estimation for the three-
    parameter . . . via the likeli-
    hood function 452
  methods of fitting a left-truncated
    . . . to discrete data 763
  of the duration of strikes 763
  three-parameter, e. g. 207, 208, 452,
    647, 735
Lognormalverteilung 41, 207, 208,
    452, 475, 503, 504, 541, 631, 647,
    735, 760, 761, 763, 770, 847, 898,
    952, 1010, 1152, 1153, 1215, 1278
Log-odds models, parallel, uniform,
    and null, e. g. 428
Log percentage change 1310
Logranktest 21, 144, 694, 770, 893,
    1291
Longitudinal:
  change, time-series models and the
    study of 954
  data:
    see also cohort analysis, and
      longitudinal studies
    analysis: bibliography 702
    graphical procedures for the
      preliminary processing of
      47
  designs, assessment and statistical
    control of subject variables in
    954
  -developmental investigations, sta-
    tistical estimation of
    structural models in 954
  models for chronic disease risk: an

    evaluation of logistic multiple
      regression and alternatives 1421
research 218, 702, 954
  causal models in 954
  rationale of, e. g. 954
studies 47, 138, 274, 416, 422, 791,
    851, 1368, 1414
  of behavioral development, the
    analysis of categorical data
    in 954
Long range dependence, e. g. 35
Longtailed symmetrical distributions,
    e. g. 1127
Lorenz curves 30
Lorenzkurve und Konzentrations-
    maß von Lorenz-Münzner bei
    Einzeldaten und bei gruppierten
    Daten, z. B. 349
Loss functions, e. g. 774
Lot-by-lot sampling inspection by
    attributes, e. g. 483
Low-dose extrapolation, confidence
    limits in, e. g. 198
LTPD-System and generalization 483
Lurking variables 121, 635

**MAD, median absolute deviation from
    the median,** e. g. 547
Mängel bei Experimenten, z. B. 1076
Mahalanobis distance, e. g. 501, 840,
    1284
  Cook's distance and Weisberg test:
    identifying outliers in regres-
    sion analysis 75, 219, 956,
    1264, 1316, 1340, 1383
  selecting the best normal
    population based on 406
Mahalanobis metric matching 1131
Mail questionnaires, e. g. 1137
  application of the randomized
    response model, e. g. 1260
Mail survey:
  nonresponse adjusting 820
  the randomized response model,
    e. g. 1260

Maintainability 265, 378, 883, 952, 999

Maintenance:
  and component reliability, e.g. 99
  schedules for failing systens 999, 1216, 1217

Mallows' C_p statistic 4, 83, 249, 300, 890, 1383

Malmquist index numbers 614

Management data base systems: see data base systems

Mandel plots, row and column 127

Mann-Whitney rank sum test: see U test

Mann-Whitney's U test: see U test

MANOVA (multivariate analysis of variance) 12, 285, 482, 587, 634, 661, 688, 716, 780, 798, 835, 840, 855, 900, 972, 1209, 1284, 1323
  see also ANOVA and MANOVA and repeated measures MANOVA, e.g. 972
  Bayesian inference in 716
  in communication research 900
  method for analyzing repeated:
    measurements, e.g. 716
    measures designs, e.g. 975
  models, nonparametric simultaneous inference for some 716
  nonorthogonal, e.g. 900
  nonparametric simultaneous inference for some models 716
  techniques, models, and applications in communication research 900

Mantel and Haenszel's Chi-square test: see Mantel-Haenszel summary odds ratio

Mantel-Haenszel:
  summary odds ratio 103, 136, 137, 367, 561, 682, 770, 1001, 1182, 1292, 1447
  test:
    see also logrank test
    for k groups by one-way analysis of variance and then F converted into chi-square 1447

Map comparison by: the coefficient of areal correspondence, the coefficient of areal correlation, the Lorenz curve, the index of dissimilarity and the pattern of differences between the maps, e.g. 1333

Maps:
  points, lines, areas and surfaces on maps, e.g. 1333
  statistical, e.g. 200, 351, 1189

Mardia test for multivariate normality, tables 712

Marginal:
  functions, testing hypotheses about 374
  homogeneity and symmetry in square contingency tables 103
  homogeneity in square contingency tables, tests for 710

Marital interactions, identifying the dominant partner in 1365

Marked point process, e.g. 233

Marker tests 2, 986, 1355

Markertest-Validierung 2, 986, 1355

Markov:
  chains 50, 51, 53, 99, 103, 193, 312, 454, 695, 768, 874, 900, 1136, 1143, 1207
  discrete . . . chains and continuous . . . processes, e.g. 99
  modelling, with additional reliability parameters such as frequency and duration, e.g. 99
  models, control theory for, e.g. 50
  process 50, 53, 99, 122, 233, 454, 552, 874, 900, 1136

Mark-recapture methods 103, 966, 1000, 1064, 1065, 1208, 1210, 1391

Marriage, probabilities of survival and dissolution of 194

Martingales, e.g. 35, 53, 454, 1136

Maschinenbau und Elektrotechnik: Beispiele zur Zuverlässigkeitsanalyse, z.B. 1130

Mass-spectra classification in chemistry 132
Matched data: analysis of 137, 367, 836, 1056
Matched pairs, e. g. 367, 836
Matched-pair Wilcoxon test: see Wilcoxon matched-pair signed rank test
Matched samples, e. g. 367
  of response curves, nonparametric analysis of two 708
Matching:
  and alternatives 21, 205, 682, 1056, 1182
  caliper, "nearest available", mean, stratified, frequency, and multivariate matching 21, 205
  confounding and overadjustment in epidemiology 135
  designs, different kinds of 21, 205
  for controlling confounding factors 21, 137, 205, 367, 682, 738, 1056, 1182, 1354
  in epidemiological studies 21, 205, 738, 868, 1182
  nearest available 1131
  pair-matching 868
    statistical analysis of pair-matched samples 21, 205, 738, 868, 1182
  versus:
    adjustment 21, 205
    stratification 21
  statistical, missing data and imputation 248, 1125
Materials testing, reliability concepts useful in, e. g. 265
Maternities 1277
Mathematical:
  models, e. g. 121, 606, 966, 1054, 1113, 1150
  tools for applied multivariate analysis, e. g. 54, 56, 443, 445, 634, 749, 814, 840, 927, 1078, 1085, 1207
Mathematics:
  and statistics: dissertation index 1331

ordering the universe 606
Mathematische:
  "Leckerbissen" 705
  Modelle, z. B. 121, 606, 966, 1054, 1150
    in der Biologie, z. B. 966, 1054, 1150
  Zeichen und Formelschreibweise 279
Mathematisches:
  Grundwissen für Biologen, z. B. 56
  Rüstzeug für multivariate Analysen, z. B. 443, 445, 1092
Mathematische (und zugleich auch musische) Knobeleien 705
Matrices:
  see also matrix algebra
  applications in multivariate analysis and linear models, e. g. 443
  circulant, e. g. 443
  dominant diagonal, e. g. 443
  Fourier matrices, e. g. 443
  Frobenius and Leslie matrices in demography, e. g. 54
  functions of, e. g. 749
  Hadamard, e. g. 54, 443, 1207
  idempotent, e. g. 54, 1207
  indicator (containing zeros and ones, with the ones indicating responses to a questionnaire), e. g. 965
  inverse positive, e. g. 443
  large data, writing programs to analyze 767
  Leslie, e. g. 54
  measuring association by correlation, covariance, and distance, e. g. 54
  multivariate, biplot display of . . . for inspection of data and diagnosis 47
  non-negative definite, e. g. 54
  orthogonal, e. g. 54, 1207
  partitioned, e. g. 54, 1207
  patterned, e. g. 443
  permutation, e. g. 443, 1207

rolled-out, e.g. 443
square matrices with nonnegative
	elements and nonnegative
	matrices with dominant
	diagonal terms and their
	application, e.g. 54
symmetric and positive definite,
	e.g. 54
theory, e.g. 54, 443, 749, 1207
Toeplitz, e.g. 54, 443
Vandermonde, e.g. 443
with positive, nonpositive, or
	nonnegative off-diagonal or
	principal minor decrements,
	e.g. 443
Matrix algebra 54, 56, 443, 445, 634,
	749, 814, 840, 927, 1078, 1085,
	1207
of regression analysis, e.g. 1207
Matrix:
arguments, hypergeometric func-
	tions of, e.g. 1078
correlation, tests for comparing
	elements of a, e.g. 1256
equations, linear, e.g. 749
hat 219, 442, 547, 548, 1264
idempotent, e.g. 54
inequalities and maximization, e.g.
	634
manipulations for correlation and
	regression, e.g. 824, 1207,
	1437
orthogonal-triangular decomposi-
	tion of a: statistical uses 661,
	1205
partitioned, e.g. 54, 1207
permanent of a, e.g. 54
polynomials and normal forms,
	e.g. 749
rank of a, e.g. 54, 443, 1207
sampling, e.g. 248
stochastic transitional, probability,
	e.g. 99
Matrizen, Matrix-algebra 54, 56, 443,
	445, 749, 814, 927, 1078, 1085,
	1207
Maximal root criterion of S. N. Roy:
	tables 712

Maximum:
absolute value of equally
	correlated normal random
	variables: tables 982
F ratio: upper 10% and 25% points
	72
likelihood:
	estimates for complete and in-
		complete two-way and
		multiway contingency
		tables 103
	estimation 11, 103, 643, 658, 769,
		774, 1390
	-Modulus-Test, z. B. 1187
McDonald und Thompson: Rang-
	summentest, z. B. 1151
McNemar-Test 470, 766, 793, 836,
	1152, 1153, 1354
	Kontinuitätskorrektur 470
	praktische Relevanz 1446
	und Gart-Test, z. B. 766
	Verallgemeinerungen 836, 837,
		1449
Mean, arithmetic, e.g. 547, 658
and other location estimators, e.g.
	327, 515, 547, 657, 774, 1248,
	1319, 1320
k-skipped 1248
largest: see there
lead time estimation, e.g. 419
random-lenght and/or fixed-
	length confidence intervals
	for the normal population
	with the largest mean 406
ridits, e.g. 611
robustness of four standard
	methods for estimating the
	989
scores, analysis of 374
selecting the population with the
	largest (smallest) mean from a
	set of k independent normal
	populations 406
square absolute error (MSAE)
	regression, e.g. 273, 928
squared errors:
	of prediction and stopping rules
		890

Mean, squared errors:
    of prediction, estimates of the, and
        the selection of subsets of re-
        gression variables, e. g. 890
    time:
        between failures, MTBF, e.g. 99,
            420, 552
        to failure, MTTF, e.g. 99, 552
        to first failure, MTTFF, e.g. 420
        to repair, MTTR, e.g. 99
        to the next failure:
            MTTTNF, e.g. 420
        nonequilibrium states and the,
            e.g. 420
Means:
    analysis of 945, 1001, 1086
    confidence regions and
        simultaneous comparisons of
        component means, e. g. 634
    contrasts of normal, simultaneous
        confidence intervals for
        monotone 838
    graphical display of 23, 200
    grouping in analysis of variance,
        e.g. 170
    linear contrasts 160, 674, 787, 1153,
        1307
    multiple comparisons of 23, 67, 68,
        73, 102, 121, 160, 203, 309,
        313, 314, 392, 393, 664, 674,
        698, 856, 895, 896, 944, 945,
        1067, 1068, 1088, 1116, 1153,
        1228, 1238, 1268, 1269, 1288,
        1448
    on partitioning . . . into groups 237
    standardization of fitted, e.g. 805
    testing the equality of several
        means when the population
        variances are unequal 102,
        203, 277, 309, 314, 393, 698,
        1116, 1288, 1397
    trimmed means 515, 547, 1248
    two-stage procedures 406
Mean:
    up time, MUT; and mean down
        time, MDT, e.g. 552
    vector, inferences about a, e.g. 634

vectors:
        likelihood ratio tests for 716
        simultaneous test procedures for
            716
Measurement:
    and survey sampling 928
    errors:
        see systematic measurement
            errors
        of 624, 733, 1081, 1199, 1387,
            1435
    in surveys, e.g. 1137
    of attitude strength 1200
    of change 422, 1320, 1321
    scales of 12, 117, 506, 663, 965, 998,
        1120, 1137, 1341, 1436
Measurements 279, 624, 663, 733,
    1081, 1137, 1387, 1435
    repeated, parametric and non-
        parametric analysis for 690
Measurement:
    theory and decisionmaking 1120
    types of: see scales of
Measures of association 9, 10, 11, 81,
    103, 210, 263, 349, 429, 611, 739,
    766, 792, 793, 798, 810, 836,
    1110, 1154, 1334
    and correlation for ordinal data,
        e.g. 10, 374, 433, 434, 439, 440,
        597, 766, 775, 792, 793, 836,
        1110
    for nominal data, e.g. 9, 103, 204,
        352, 353, 355, 425–429, 792,
        836, 1110, 1334
    for nominal-ordinal data, e.g. 9,
        103, 204, 352, 353, 355, 367,
        425–429, 766, 792, 836, 1110,
        1334
Measures:
    of inequality and Lorenz curves 30
    of location and dispersion, e.g. 658
    repeated, design and analysis of
        31, 44, 322, 516, 836, 856, 895,
        896, 901, 1017
Measuring:
    change 103, 422, 1320, 1321
    methods, comparison of several
        172

64

relative change of ratio-scale
variables 1310
similarity and dissimilarity: see
cluster analysis
Median:
absolute deviation from the
median, e. g. 547
and hinge traces, e. g. 502
and its standard error 1151–1153
k-times folded without and with
trimming 1248
polish 547, 871, 916, 1320, 1339
and analysis by means 547
and least-absolute-residuals fit
547
resistance of 547
test, multivariate, e. g. 708
und zugehöriger Standardfehler
1151–1153
Medical:
diagnosis 78, 79, 86, 132, 250, 390,
986, 1355
follow-up studies 96, 194, 682, 694,
1314, 1348
see also cohort studies
informatics, e. g. 620
screening 79, 225, 367, 419, 1145
trials: see clinical trials
Medien: Daten 584
Medizinische:
Daten: Zusammenhangsanalysen,
z. B. 1384
Kommunikationsabläufe und ihre
Darstellung durch Petri-
Netze 86, 692
Versorgung der Bevölkerung der
Bundesrepublik Deutschland
1079, 1255
Medizin, Prognose- und Entscheidungsfindung in der 620
Mehrdimensionale:
Skalierung 54, 83, 238, 257, 346,
435, 446, 482, 501, 603, 612,
634, 717, 840, 969, 1078, 1195,
1414, 1436
Varianzanalyse: siehe MANOVA
Mehrfeldertafel, siehe Kontingenz-
tafel

Mehrkörpersysteme, Dynamik von,
Simulation der, z. B. 424
Mehrphasige und mehrstufige Stichproben, z. B. 779
Meinungsumfragen: siehe Erhebungen
Melatonin levels for 11 subjects over
13 months: averaging procedure
and fitting a standard curve 241
Messungen: Theorie des Messens,
Skalierungen 12, 117, 506, 733,
998, 1120, 1341, 1436
Messy data, analysis of, e. g. 896
Meßfehler, systematische und zufällige 624, 733, 1081, 1199, 1387,
1435
Meßtechnik 279, 733, 1081
siehe auch Kalibrierung
M- and L- estimators and outliers,
e. g. 71
M-estimators, e. g. 547, 719, 853
Meta-analysis, e. g. 414, 588, 1132,
1270
evaluation of, e. g. 414
jacknife technique, e. g. 414
Meta-analytic procedures in social
research, e. g. 414, 588, 1132,
1270
Methodbase systems, e. g. 507
Method-comparison data, statistical
evaluation of 172, 1427
Method development and evaluation, criteria for judging precision and accuracy in 279, 733,
1081, 1387
Methodenvergleich: enzymatische
Aktivitäten 172
Methodenvergleiche in der klinischen Chemie, z. B. 1427
Method validating in analytical
chemistry 172, 289, 733, 1387
Metrology: propagation of error
formulas 733
Microcomputer:
as the statistician's mainframe 925
regression program, e. g. 15
statistical packages for 171, 404,
520, 679, 741, 918, 933, 1423

Migration in human population 37, 161, 340, 455, 591, 818, 1148, 1255, 1281

Mikrozensuserhebungen 133, 161

Minimal cut:
  and path sets 99, 524
  set method: identification of critical areas of system weakness, e.g. 99, 524

Minimale Stichprobenumfänge für Varianzanalysen 461, 462, 1045

Minimaxity, e.g. 87, 774

Minimax regret sampling plans 483

Minimization, sequential unconstrained, e.g. 860

Minimum:
  absolute deviation (MAD) regression 273, 929
  chi-square estimation and polytomous models, e.g. 13
  distance procedures 719
  sum of absolute errors regression: a state of the art survey 273, 929
  variance estimation, e.g. 547, 658, 1072

Mining: sampling of particulate materials 469

Minitab: EDA, e.g. 1339

MINQUE estimation: see variance components

Mischungsproblem: Ermittlung der optimalen Zusammensetzung 224, 333, 1257

Mischverteilungen 149, 224, 307, 343, 869, 1127

Misclassification bias in epidemiology 450

Missing data:
  in agricultural field experiment 1036
  estimating: by using prior knowledge, inserting mean values or using regression to predict missing values, e.g. 1284
  in surveys 804, 820, 1125, 1137, 1424

Missing observations:
  creating a dummy variable, e.g. 1284
  how to cope with 538, 732, 896, 1284

Missing:
  or spoilt data, e.g. 1036
  values:
    how to deal with in multiple regression analysis 210, 295
    inference and data tables with 717

Mittelwerte:
  graphisch dargestellt 23, 200
  ihre Zerlegung in Gruppen, z. B. 237
  Vergleich mit einer Kontrolle 68, 369, 466, 787

Mittelwertvergleiche: Gleichheit mehrerer Mittelwerte bei ungleichen Varianzen 102, 203, 277, 393

MIVQUE estimation: see variance components

Mixed:
  binomial distribution 483
  model analysis versus MANOVA approach for repeated measures, e.g. 975
  models 31, 264, 896, 1205
  normal distributions 149, 224, 343

Mixture:
  data, analysis of 224, 333, 1257
  design 224, 1257
  designs with multiple constrains on the component proportions 224
  experimentation using the ratios of their components, e.g. 224
  distributions 149, 224, 307, 343, 869, 1127
  experiments 224, 1257
    see blending, fish patties, and fruit punch mixture
    and extreme vertices designs 1257
  models in experiments 224

Modelle:
erste Einführung z. B. 121, 615, 656, 1150
mathematische und statistische ... in Biologie und Medizin 615, 692, 723, 966, 1054
Modellierung dynamischen Verhaltens komplexer sozio-technischer Systeme durch Petri-Netze 410, 529, 1104, 1105
Modelling:
checklist, e. g. 615
mathematical, e. g. 121, 606, 966, 1054, 1113, 1150
statistical, an introduction to, e. g. 288
Model selection 1046
Models:
and simulation: see simulation and Monte Carlo method
Bayesian analysis of linear 143
decision-aiding 398, 654, 1119, 1120
for policy analysis 398
discriminating among rival 165, 1046
distributional, for data, exploring, graphical methods, e. g. 417
fitting additive, to two-way data 1142
for cumulative logits, e. g. 11
for describing software reliability, e. g. 46
for duration and size of social processes 50
for experiments with mixtures 224
formulation and adequacy 25, 121, 288, 300, 398, 474, 615, 656, 822, 829, 830, 864, 1001, 1150, 1215, 1257, 1369, 1383
for nonresponse in sample surveys 205, 248, 594, 714, 804, 820, 928, 1153, 1161, 1275
for social systems:
and processes, e. g. 50
with fixed class sizes, e. g. 50
for survival data, e. g. 193, 234, 332, 770, 864, 952

for the analysis of associations in cross-classifications having ordered categories 10, 204, 263, 426–428, 1027, 1357
for the diffusion of news, rumours and ideas, e. g. 50
for the duration of strikes 763
for time to failure of various types of objects: see survival data analysis, lognormal distribution and Weibull distribution
generalized linear, e. g. 34, 288, 409, 863, 864
in longitudinal research 954
in statistics: see also distributions
interactive, e. g. 1113
in the practice of survey sampling 644
introduction to, e. g. 121, 615, 656, 1150
linear: see linear models
loglinear 21, 42, 103, 288, 346, 352, 353, 355, 425, 428, 472, 513, 564, 798, 822, 862, 864, 888, 1110, 1384
goodness-of-fit procedures for the evaluation and selection of 115
software, e. g. 114, 285, 288, 409, 428, 471, 472, 863, 864, 472
mathematical 121, 966, 1054, 1150
multiplicative and cohort analysis 138
non-linear regression: see non-linear regression models
of arrival processes, e. g. 760
of change: see Panel Analyse
of conflict: see game theory
of growth 21, 44, 178, 300, 422, 566, 690, 703, 1065
of matching 21, 205
of multidimensional population dynamics 750
of progressive diseases 419
spatial statistics 389, 968, 1118, 1333

statistical, in biological and medical sciences 615, 692, 723, 966, 1054

structural equation models 169, 220, 358, 926, 1404

unobserved variables within 358

thinking with, e.g. 1150

unbalanced:

data, fitting analysis of variance models, e.g. 780, 789, 1043, 1142, 1259

two-way data, fitting additive models to, e.g. 1142

Moderne Statistik 499, 532

Modes and isopleths, estimating both 1155

Moment inequalities 1028

Moments:

and cumulants 643, 658, 1028

cumulants and generating functions for 20 continuous and 6 discrete distributions 1028

of order statistics from some discrete and continuous distributions 1028

Monetary-unit sampling, e.g. 1395

Monte Carlo:

integration and variance reduction techniques, e.g. 1143

method: see also simulation

optimization, e.g. 1143

simulation:

see simulation

output data, statistical analysis of, e.g. 759

uniform repair distribution and modified normal distribution, e.g. 420

Morbidität und Mortalität 6, 86, 133, 137, 161, 194, 260, 332, 418, 591, 682, 738, 742, 753, 791, 797, 1079, 1140, 1230, 1255

Morphometrics, multivariate 717

Mortalitätsraten 161, 194, 340, 602, 1148, 1428

Mortality analysis 133, 166, 194, 225, 260, 282, 332, 340, 682, 791, 864, 888, 1230, 1321, 1428

in time 166, 282, 340, 818, 1428

manual of 166

Mortality:

APC models of 851

differential mortality 166

index and relative mortality index with their standard errors 194

indices 166, 194, 260, 340, 602, 1428

infant and maternal 166, 194

in heterogeneous population 668

laws, e.g. 194

odds ratio and proportional mortality ratios 602

ratio:

proportionate . . . and standardized mortality ratio, a comparison 260

standardized, and related adjusted rates with their standard errors 194

survival and density estimates for animal populations 1208

trends 166, 340, 818, 1428

Moving average:

see also forecasting and time series analysis

processes, e.g. 4, 122

Moving averages deceptiveness of 942

MTBF, MTTFF and MTTTNF, e.g. 420

Müttersterblichkeit 1079, 1255

Multiattribute evaluation, e.g. 324

Multi-center controlled trials, e.g. 112

Multicollinearity 4, 54, 83, 446, 902, 955, 1316, 1340, 1350, 1402, 1420, 1437

and ridge regression 555, 956, 1015, 1350, 1385

and singularity: remedies e.g. 1284

measures, meaningful, e.g. 1402

Multidimensional:

contingency tables: see multiway (contingency) tables

data analysis and quantification of categorical variables, e.g. 612

Multidimensional:
  scaling 54, 83, 238, 257, 346, 435,
      446, 482, 501, 603, 612, 634,
      717, 840, 969, 1078, 1195,
      1414, 1436
    nonmetric 435, 969
  sorting, e. g. 97
  unfolding and individual scaling
      435
Multilevel experiments with qualita-
    tive and quantitative variables
    270
Multilinear models, e. g. 417
Multinomial distribution 103, 352
  and selection problems 406
  bounded 747, 748
  curtailment problem for the 406
Multinomialverteilung und Bayes-
    Verfahren 498
Multinormal distribution 31, 446,
    658, 799, 800, 840, 1078, 1209
Multi-phase sampling, e. g. 621
Multiple:
  and partial correlation and regres-
      sion 210, 300, 349, 663, 798,
      831, 835, 899, 955, 1383, 1390,
      1420
  comparisons:
    see also multiple Vergleiche
    and orthogonal contrasts, e. g.
        67, 83, 286, 789, 1408
    and type III errors 111
    error rates in 160, 674, 1448
    Holm procedures 565, 1180, 1242
    in sample proportions for tests
        of homogeneity 836
    of correlation coefficients: a
        robust test 317
    of means 23, 67, 68, 73, 83, 102,
        111, 121, 160, 309, 313, 314,
        392, 393, 467, 664, 674, 698,
        856, 895, 896, 944, 945,
        1067, 1068, 1088, 1116,
        1153, 1228, 1238, 1268,
        1269, 1288, 1448
      unknown and unequal
          variances 102, 203, 277,
          306, 309, 314, 393, 698,
          1116, 1288, 1397
      on the theory for Ryan's univer-
          sal ... procedure with treat-
          ment of ties in the ranks of
          samples 570
      raising the α level in proportion
          to the judgment of the im-
          portance of a particular
          contrast 1133
      robust 102, 203, 306, 309, 314,
          393, 698, 1116, 1288, 1397
      with a control, unknown
          variances, new tables for
          306, 309
  correlation 8, 210, 658, 771, 798,
      840
    coefficient: tables of upper per-
        centage points 771
    squared, adjusted, e. g. 1284
  covariance analysis 585
  discriminant analysis: see discrimi-
      nant (function) analysis
  linear regression 8, 83, 93, 114, 206,
      210, 273, 285, 288, 295, 300,
      349, 482, 663, 780, 798, 824,
      831, 835, 890, 899, 902, 929,
      955, 956, 1001, 1238, 1284,
      1383, 1390, 1420
  regression analysis 8, 83, 93, 114,
      206, 210, 285, 288, 295, 300,
      349, 482, 663, 780, 798, 824,
      831, 835, 890, 899, 902, 955,
      956, 1001, 1238, 1284, 1383,
      1390, 1420
    choice of subsets in, e. g. 300,
        549, 550, 890
    dealing with missing values 295
    estimating optimal transforma-
        tions, e. g. 97
    influence-enhanced scatter plots
        1296
    linear equality restrictions in
        1115
    selection of subsets of regression
        variables, e. g. 300, 549, 550,
        890

regression and time series analysis, e. g. 1420

Regression mit Prozentzahlen, z. B. 799

Regressionsanalyse 8, 83, 210, 300, 349, 482, 663, 780, 798, 831, 835, 899, 902, 955, 1383, 1390, 1420

regression:
  under linear restrictions, e. g. 661, 1115
  univariate and multivariate, e. g. 899
  with categorical variables, e. g. 1110

sampling, e. g. 1181

time series model building, e. g. 958

und partielle Korrelationsanalyse 210, 798

Vergleiche:
  siehe auch multiple comparisons
  Komplexität 1068
  Methode nach Ryan mit Bindungen 570
  mittels Rangtests: alle Paarvergleiche 567
  nach Holm 565, 1180, 1242
  nach Tukey 664, 1088
  robuste bzw. bei ungleichen Varianzen 102, 203, 309, 314, 393, 698, 1116, 1288, 1397
  und Holm-Prozeduren 565, 1180, 1242
  Verallgemeinerung der Dunnett-Prozedur 1067
  von Mittelwerten: 23, 67, 68, 73, 102, 121, 160, 203, 309, 313, 314, 392, 393, 664, 674, 698, 856, 895, 896, 944, 945, 1067, 1068, 1088, 1116, 1153, 1228, 1238, 1268, 1269, 1288, 1448

Multipler Korrelationskoeffizient: obere Schranken 771

Multiplicative:
  log-linear hazards model 234

models:
  and cohort analysis 138
  for rates and methods for analyzing categoric and censored survival data 562

Multiplikative Modelle 219, 562, 1122
  loglineare Modelle für multinomial-verteilte Merkmale und Kovarianzselektion für normalverteilte Merkmale 1384

Multiplicative odds model or logit model, e. g. 425

Multiplikatives loglineares Hazardmodell 219

Multiseriale Korrelation 739

Multiserver queues, tables of waiting times 1014

Multistage sampling 621, 645, 729, 781, 1137, 1274, 1433

Multistate systems, e. g. 265

Multi-stratification 621

Multivariable:
  experiments, two-level 270
  functions, symbolic differentiation and optimization of, e. g. 445

Multivariate analysis 8, 31, 342, 346, 350, 370, 446, 482, 501, 513, 612, 634, 650, 712, 752, 800, 814, 835, 840, 855, 899, 900, 992, 1078, 1103, 1195, 1209, 1284, 1320, 1323, 1366, 1384, 1399
  of contingency tables 103, 352, 353, 355, 835, 1110, 1334
  of covariance 285, 585, 1209, 1284
  of cross-classifications, e. g. 1110
  of variance 12, 285, 482, 587, 634, 661, 688, 716, 780, 798, 835, 840, 855, 900, 972, 1209, 1284, 1323

Multivariate:
  Bayesian analysis, e. g. 1078
  calibration 148
  continuous distributions, e. g. 631, 712
  correction for restriction in range, e. g. 798

Multivariate:
  data:
    analysis of: see multivariate
        methods
    displaying 47, 69, 70, 97, 183,
        342, 351, 370, 634, 683,
        1209, 1317, 1319, 1359, 1366
    graphical displays and methods
        47, 69, 70, 97, 183, 342, 351,
        370, 634, 683, 1209, 1317,
        1319, 1359, 1366
  descriptive statistical analysis:
      correspondence analysis 47,
      164, 448, 535, 596, 608, 767,
      800, 965
  discrete distributions 103, 632, 1030
  distributions 22, 30, 103, 188, 632,
      712, 716, 814, 1030, 1038,
      1209, 1215
    computation of some 716
  linear models, e.g. 31, 288, 1078,
      1209
    Bayesian analysis of 143
  logistic analysis of risk factors 137,
      682, 770, 1182, 1314
  matched sampling: constructing a
      control group 1131
  means and linear models, infer-
      ences about, e.g. 634
  median and sign tests, e.g. 708
  methods 8, 31, 249, 342, 346, 350,
      370, 445, 446, 482, 501, 513,
      612, 634, 650, 683, 712, 752,
      800, 814, 824, 835, 840, 855,
      899, 900, 992, 1078, 1103,
      1195, 1209, 1284, 1320, 1323,
      1359, 1366, 1384, 1399
    a list of computer programs, e.g.
        1078
    erratic data points and breaking
        down of the theoretical
        assumptions, e.g. 634
    in ecological work 997
    statistical tables for, e.g. 712,
        1038, 1209
    visual display of multivariate
        data e.g. 342, 351, 370,
        1209, 1317, 1359, 1366

morphometrics 717
multiple regression versus
    MANOVA and canonical
    correlation 899
noncentral t distribution: power
    curves for the analysis of
    means 946
non-parametric tolerance regions:
    sample size 5
normal distribution 22, 31, 446,
    634, 658, 799, 800, 840, 1078,
    1209
  tables related to the 712
normal distributions, selection
    procedures for 406
normality:
  see also there
  assessing 22, 342, 634
  robust methods in the assess-
      ment of 853
normal samples 814
Normalverteilung 11, 31, 446, 634,
    658, 799, 800, 840, 1078, 1209
  und klinisch-chemische Daten
      1103
observations 8, 31, 274, 342, 346,
    350, 370, 446, 482, 501, 513,
    612, 650, 683, 712, 752, 800,
    814, 835, 840, 855, 899, 900,
    992, 1078, 1103, 1195, 1209,
    1284, 1320, 1323, 1359, 1366,
    1384, 1399
Pareto distribution 30
path models 1277
populations, structuring 1078
problems: tables 712, 1038
regression analysis, e.g. 840, 1368
  see also multiple regression
regression, assessing
    dimensionality in 716
selection problems 406
statistical distributions:
  see also multivariate distribu-
      tions
  tables 22, 712, 814, 1038
statistical hypotheses, tables for
    testing 712, 814, 1038

statistical packages: a comparison 900

structure, Bayesian approaches to 47

t-Verteilung 188

Multiway (contingency) tables:

analysis of 11, 103, 204, 285, 323, 334, 335, 352, 353, 355, 426–429, 472, 513, 835, 1056, 1334, 1384

saturated model and unsaturated models, e. g. 1334

Muscle biopsies data with 11 histological variables scored as 0, 1, 2, 3: allocation to one of two populations 537

Muster-Erkennung 490, 612, 962–964, 1083, 1085

Muster-Stichprobenpläne Deutscher Marktforschungsinstitute 1169

MUT (mean up time), e. g. 552

Mutagen screening 1313

Mutationshäufigkeiten 1313

Nachnamen: welche Anfangsbuchstaben sind eher typisch deutsch? 1166

Nachrichtenübermittlung, geheime 91, 92, 187, 606, 696

Namensstichproben 1166

Narrow limit gauging, e. g. 1181

Natalität, Geburtenhäufigkeit 340, 1428

National:

income accounts 1040

statistics:

FRG 1255

UK 1106

USA 671

Natural selection and mutation 331

Nearest:

available matching 1131

neighbour:

distance, mean, e. g. 1333

methods, e. g. 490, 717

Negative binomial distribution 103, 298, 504, 631, 816, 1028, 1065, 1124, 1145

table predicting the length of series until at least *k* events have taken place given the probability *p* of this event 1145

Negative:

Binomialverteilung 103, 298, 504, 542, 631, 816, 1124, 1145

multinomial distribution 103

Negotiations, two person, power and 128

Neighbour designs 40

Nested:

data: fitting equations 249

designs 123, 674, 896, 1238, 1239, 1408

factors, nested design, hierarchical classifications 123, 674, 896, 1239, 1408

Nets:

consisting of conditions and events, systems organization using 410, 529, 1016, 1104, 1105

Petri 410, 529, 1016, 1104, 1105

with individual tokens, e. g. 1105

Network:

algorithms 1315

analysis methods, communication 900

and connectivity matrix, e. g. 1333

data model 1330

measures on, e. g. 1333

model and the proposal of the data bank task group of the conference on data systems languages 1330

modelling: simple and complex systems, e. g. 99

reliability, e. g. 46, 97

Networks:

topologically distinct channel, e. g. 1333

tree, e. g. 1333

Netztheorie, Theorie der Systemorganisation 410, 529, 1104, 1105

New better (worse) than used distributions, e. g. 1028

Newman-Keuls-Student-Test 73, 160, 313, 674, 732, 1017, 1088

News, ideas and rumours: diffusion models 50, 51, 53, 966

Newton-Cotes quadrature, e. g. 661

Newton-Raphson algorithm, e. g. 472, 770

Newton's method:
  and quasi-Newton methods, e. g. 661, 860
  for the nonlinear regression problem, e. g. 661
  for unconstrained minimization, and some modifications, e. g. 661

Neyman's contagious distributions, e. g. 298, 542

Neyman-Scott process and Bartlett-Lewis process, e. g. 233

Neyman Typ A Verteilung 298, 542

Nichtadditivität: ihre Beseitigung bei Zweiwegklassifikation mit einer Beobachtung 240

Nichtlineare:
  Kalibrierfunktionen: see calibration
  kleinste Quadrate 219, 249, 300
  Programmierung, z. B. 333, 860
  und dynamische Programmierung: siehe Operations Research
  Regression 8, 249, 300, 587, 651, 824, 956, 1238

Nichtmetrische multidimensionale Skalierung 969

Nichtnormalverteilung, Prüfung auf 129, 290, 547, 716, 806, 853, 1152, 1153, 1390

Nichtorthogonale:
  Daten und lineare Modelle 937
  Varianzanalyse 239, 780

Nichtparametrische:
  Regression 590
  Statistik 214, 215, 312, 315, 363, 369, 703, 719, 770, 793, 836, 893, 1024, 1072, 1090, 1225, 1288

Tests auf Streuungsunterschiede 215, 315, 1044

Nichtzentrale:
  χ^2-Verteilungen 514, 541, 1028, 1032
  F-Verteilungen 462, 489, 525, 541, 970, 1028, 1032
  t-Verteilungen 541, 633, 1005, 1028
  Verteilungen: siehe noncentral distributions

Nodality, mean, e. g. 1333

Nominal:
  data indices of variation, e. g. 1110
  measures of association 9, 103, 204, 352, 353, 355, 425–429, 792, 836, 1110, 1334

Nonadditivitiy: Tukey-test, e. g. 240, 674, 732

Noncentral:
  chi-squared distribution 514, 541, 1028, 1032
  distributions 462, 489, 514, 525, 541, 633, 970, 1005, 1028, 1032
  see also the special distribution
  F-distribution 462, 489, 525, 541, 970, 1028, 1032
  t-distribution 541, 633, 1005, 1028

Nondestructive testing, statistical analyses for, e. g. 265

Nonequivalent:
  group designs, e. g. 220
  groups pretest–posttests designs, e. g. 585

Nonexperimental:
  causal inference and structural models, e. g. 1404
  designs in the social sciences: an introduction, e. g. 1246
  studies: see observational studies

Nonleast squares regression: see ridge regression and robust regression

Non-Gaussian symmetrical long-tailed distributions 1127

Nonlinear:
  analysis of covariance, e. g. 585
  covariance analysis, e. g. 585

estimation in regression analysis:
see regression, nonlinear
Nonlinearity, e.g. 1098
Nonlinear:
  least squares 219, 249, 300
  least squares, the geometry of 300
  mapping, e.g. 342
  methods in statistical computation,
    e.g. 824
  programming, e.g. 536, 860
  regression models 8, 249, 85, 300,
    587, 651, 661, 732, 824, 956,
    1082, 1098, 1238
    estimation of parameters in,
      e.g. 716, 1098
  least absolute value estimation in
    275
  time series models, e.g. 958
NONLINWOOD 249
NONMET: see GSK-Ansatz
Nonmetric multidimensional scaling
  342, 435, 513, 840, 969
Nonnormality:
  kinds of 630, 915, 1127, 1153, 1427
  and variance heterogeneity inter-
    action, e.g. 126
  due to skewness and kurtosis, e.g.
    1153
  testing 129, 290, 781, 806, 821,
    1152, 1153
Nonorthogonal:
  analysis of variance 239, 780, 900
  data and linear models, e.g. 937
  MANOVA, e.g. 900
Nonparametric analysis:
  of growth and response curves, e.g.
    703
  of pairs: methods and programs
    1178
  of response curves 703, 708
Nonparametric:
  Bayesian procedures, e.g. 498
  calibration 685
  classification, e.g. 717
  density estimation 63, 1024, 1025,
    1289, 1386
  discrimination, e.g. 1071

functional estimation 1071
integration, e.g. 414
multivariate tolerance regions:
  designing the sample size for 5
point estimation, e.g. 1090
regression 201, 590
sequential estimation 719
statistical data modeling, e.g. 547,
  719, 1024, 1025
statistics, e.g. 214, 215, 312, 315,
  363, 369, 701, 719, 770, 793,
  836, 893, 1024, 1072, 1090,
  1225, 1288
  a survey of Bayesian approaches
    to 1225
  graphical methods, e.g. 363
  selected tables for 719
  theory of, e.g. 1072, 1090
survival data analysis, e.g. 770, 893
testing:
  for location and scale para-
    meters 214, 782, 793
  for scale 214, 215, 315, 793
  for time series, e.g. 312
  for variances 214, 215, 315, 782
  tests of directional data, e.g. 839
Nonproportional hazards, e.g. 173,
  437
Nonrandom samples, on the validity
  of inferences from, e.g. 1237
Nonresponse:
  and imputation 248, 714, 820, 1161
  and missing values 248, 714, 820,
    1161, 1237
  in surveys 205, 422, 594, 645, 804,
    820, 928, 1153, 1275
Nonrobustness effects, e.g. 126, 203
Nonsampling errors in surveys 594,
  714, 727, 1424
Non-stationary time series, e.g. 122,
  720, 1019
Non-uniform random variables
  generation, e.g. 908
Normal:
  correlation model, e.g. 210, 956
  deviates, generating pseudo-
    random 421, 661, 1035

Normal:
  distribution 643, 661, 1029, 1392
    and empirical Bayes estimation,
      e.g. 175
    and selection problems, e.g. 406
    approximations 39, 661, 1029
    contaminated 149
    deviations from 441, 630, 915,
      1127, 1427
    generation of variates from the
      tails, e.g. 1186
    models: univariate and multi-
      variate models, e.g. 31
    modified: a special time-to-fail-
      ure distribution, e.g. 420
    multivariate 31, 446, 634, 658,
      799, 800, 840, 1078, 1209
    percentage points of the 65, 661,
      1392
    probabilities 39, 661, 1029
    spherical 31, 1373
    tables of normal percentile
      points 1392
    testing normal distributional
      assumptions by using
      probability plots 286, 821,
      1153, 1238
    truncated, a graphical method
      for estimating the parame-
      ters of a 195
  distributions:
    contaminated, arising in clinical
      chemistry, estimation
      procedures 149
    mixtures of 343
  equations for the more important
    functional relations, e.g. 1153
Normalformen von Relationen 1349
Normal forms:
  in method bases, e.g. 507
  of tables: see normalization of
    tables
Normality:
  assumption, e.g. 126
  bivariate, evaluating, e.g. 634
  multivariate 22, 31, 446, 634, 658,
    799, 800, 840, 1052, 1078, 1209

  robust methods in the assess-
    ment of 853
  tables for testing 712
  testing 22, 634, 716, 1052
testing by plot 290, 806, 821
tests:
  for 129, 290, 547, 716, 806, 853,
    1152, 1153, 1390
  of univariate and multivariate
    634, 716
Normalization of tables 576, 1330, 1349
Normal:
  means: simultaneous confidence
    intervals for monotone
    contrasts 838
  mixtures 149, 224, 343
  -normal and beta-normal
    mixtures: iterative percentile
    versus maximum likelihood
    estimation 149
  percentile points, tables of the 1392
  plot:
    see also normal probability plot
      and half-normal plot, e.g. 300
      or rankit plot, e.g. 300, 821,
      1383, 1390
  populations:
    a table for interval estimation of
      the largest mean of k 189
    interval estimation of the largest
      mean of k 189
  probabilities, a simple approxima-
    tion for bivariate 875
  probability plots 286, 300, 808, 821,
    1052, 1153, 1238, 1383
    see also probability plots
    use of the correlation coefficient
      with 806
  random numbers, generation of,
    e.g. 421, 1035
  scores, e.g. 836
  tail approximating 510
  tolerance:
    limits: tables 921, 983
    regions, improved bivariate 485
  variables, correlated, testing for
    equality of variances of 512

Normalverteilung 643, 1029, 1392
  Abweichungen 630, 915, 1127,
    1153, 1427
    von einer ... bei 20 Serumbe-
      standteilen 1427
  bivariate 643, 798, 875, 1008, 1029
  mit bekanntem Variations-
    koeffizienten, Schätz- und
    Testverfahren bei 267
  Varianz, Tabellen 802
Normen für Größen und Einheiten
  in Naturwissenschaft und
  Technik 279
Nuclear magnetic resonance: appli-
  cation of factor analysis 827
Nullbefund in der Medizin und sein
  Vorhersagegewinn 692
Null log-odds model, e.g. 428
Nullwachstum einer menschlichen
  Bevölkerung: Reproduktions-
  masse und demographischer
  Lastenquotient 505
Number-of-errors distributions, dis-
  tances between and orderings of
  659
Numerical computation, difficulties
  in 520
Nuptialität, Heirats- und Ehestatistik
  340, 1428
Nutrient cycling, e.g. 1065

O, o, O_p, o_p notation (asymptotic
  methods), e.g. 103
Observational studies, e.g. 21, 205,
  791, 1131
OC functions: of single, multiple and
  sequential sampling plans, e.g.
  483
Occupational:
  mobility inter- and
    intragenerational 50
  studies, e.g. 225
Odds:
  and odds-changing evidence, e.g.
    915

and risk ratios: Bayesian inference
  569
log-linear models for odds: the
  analysis of dependence in
  cross-classifications having
  ordered categories 428
multiplicative odds model or logit
  model, e.g. 425
on death, logarithm of 437
ratio 6, 10, 11, 21, 86, 135, 137, 352,
  367, 396, 397, 450, 472, 561,
  569, 574, 682, 888, 1110, 1182,
  1292, 1299, 1447
  adjusting using auxiliary infor-
    mation in order to estimate
    the relative risk 561
  approximate confidence limits
    397, 1182
  conditions for confounding of
    the 113
  estimation: matched analysis to
    estimate the odds ratio with
    at least one control per case
    1182
  in $2 \cdot c$ tables 574
  in $r \cdot c$ tables 425, 428, 429, 1110
Odeh Test, z. B. 1151
Ökonometrie, z. B. 650
Official statistics:
  international trends in 35
  United Kingdom 1106
Omission bias in best subset selection
  890
One-sample:
  location problem, e.g. 1090
  symmetry procedures in nonpara-
    metric graphical statistics 363
Operations Research 156, 310, 333,
  366, 491, 536, 695, 885, 1136,
  1194
  survival analysis in 974
Opinions aggregation, e.g. 612
Opinion research, see survey and
Optimal:
  continuous sampling procedures
    781
  experimental design, e.g. 33, 1257

Optimal:
    the general equivalence theorem phrased in terms of design measure 1257
    scaling and alternating least squares, the combination of, e.g. 1436

Optimale Schichtungen in der Stich-probentheorie 266, 302, 1191

Optimal inspection for failing systems 1216, 1217

Optimality criteria in experimental design, e.g. 120, 1257

Optimal:
    process control by gauging 781
    skip-lot single sampling plans for Markov chains 781

Optimization constrained, e.g. 556

Oral delivery, visual aids and hand-outs 379

Order effects, Gart's exact test for, e.g. 766

Ordered:
    categories: association in cross-classifications 10, 352, 353, 426–428, 861, 1027, 1357
    contrasts: ensemble-adjusted p values 1133
    random variables and their functions 253, 495, 658

Order:
    (accuracy) of an approximation, e.g. 103
    statistics 214, 253, 495, 658, 793, 836, 1028, 1209
        and empirical distribution 719
        optimal for location and scale parameter estimation, e.g. 253, 547, 1024, 1025
        tables for 719

Ordinal:
    categorial data, analysis of, e.g. 11
    correlation coefficients, e.g. 1110
    data:
        analysis of, e.g. 11
        regression 11, 18, 862
    measures of association and correlation 10, 11, 374, 433, 434, 439, 440, 597, 766, 775, 792, 793, 836, 1110

Ordinalskala:
    -Assoziationsmaße, z.B. 11, 766
    und Kreuzklassifikation 10, 204
    und Regression, z.B. 11, 18, 862

Ordination or mapping techniques to obtain a representation of the multi-dimensional original data in a reduced number of dimensions, e.g. 342, 769, 810

Ordinationsproblem 596

Ordination techniques: direct, polar, and principal components ordination, e.g. 1065

Organizations producing statistics 846

Organoleptik 618, 1107

Orientation data statistics 57, 364, 839, 840, 1029, 1373

Ornithology, statistics in 909

Orthogonal:
    coefficients for unequal intervals and unequal n's 674
    contrasts: multiple comparisons, e.g. 67, 83, 286, 789, 955, 1408

Orthogonality in experimental designs, e.g. 1036

Orthogonalization:
    techniques for the analysis of linear models 661, 824, 1205
    -triangularization methods in statistical computation, e.g. 661, 824, 1205

Orthogonal polynomials, e.g. 300, 661, 732, 930, 931
    unequal spacing and frequencies 930

OSIRIS III: see e.g. factor analysis

Ott-Free Test, z.B. 1151

Outliers:
    detecting and reducing the influence of outliers in unvariate, bivariate, and multivariate data, and in designed experiments 49, 71, 75, 253, 350, 511, 515, 547, 607, 733, 1135, 1264, 1424

see also:
    hat matrix
    influential observations in
        regression analysis
    robustness
    Studentized residuals
detecting:
    power and some other consider-
        ations 607
    up to 10 outliers in samples of
        $25 \le n \le 500$, tables 1135
    extreme Studentized deviate
        (ESD) many-outlier proce-
        dure, a generalized 1135
    multivariate, looking at ... from the
        consultant statistician's point
        of view 350
Outputkontrollen zur Sicherung sta-
    tistischer Datenbanken, z. B.
    1185
Oval plots to show cyclic behavior
    1319
Overall and multiple hypothesis
    testing, e. g. 59
Overlapping peaks or wave forms,
    resolution of 377

Paardifferenzen-Homogenitätstests,
    verteilungsfreie 708
Paarvergleiche:
    Dominanz- und Ähnlichkeits-,
        z. B. 117
    multiple, mittels Rangtests 567
Packages, conversational, in statisti-
    cal computing, e. g. 7
Page test, e. g. 543, 1152, 1153
    exact distribution 979
Paired:
    comparisons:
        a bibliography on the method of
            254
        and a repeated measures design,
            e. g. 634
        basic procedures and examples
            719

observations:
    comparing proportions, e. g.
        1072
    comparing treatment-induced
        changes for k independent
        samples 795
    studying, by graphical methods
        183
samples:
    comparing dispersion 1162
    distribution-free tests of
        homogeneity 708
    t-test 1075
    t-test: generalization 1134
    versus two-sample design, e. g.
        1354
    for dichotomous outcome 1354
Pair-matched samples, statistical
    analysis of 21, 205, 738, 868
Pair-matching 21, 205, 682, 738, 868
Pairs, nonparametric analysis of,
    methods and programs for 1178
Pairwise:
    comparisons:
        following Waller and Duncan,
            e. g. 1001
        simultaneous, representations
            of, e. g. 716
    multiple comparison procedures
        for parametric and nonpara-
        metric analysis of variance
        models 1448
Palm distribution, e. g. 233
Panel:
    analysis/Panel-Analyse 36, 103,
        274, 425, 645, 665, 842, 1274,
        1334
    correlation, crosslagged, e. g. 662
    data, analyzing: see panel analysis
Paper, presenting a statistical 328,
    379, 1263
Paradoxien und Antinomien 882, 884
PARAFAC preprocessing 417
Parallel:
    curve assays 160, 359, 360, 531, 582
    line assays 160, 359, 531, 582, 1095
    log-odds model, e. g. 428

Parameter:
see estimation theory and test
theory
Schätz- und Testtheorie für: siehe
Schätztheorie und test theory
statistische Inferenz für lineare,
z. B. 587, 658
Pareto distributions 30, 504, 541, 613,
631, 1028
Parity progression ratio for a current
population, e. g. 194
Partial:
and multiple correlation and
regression, e. g. 93, 658, 1238,
1437
autocorrelation, e. g. 4, 122
leverage regression plots 75, 219
regression plots, e. g. 1316
residual plot and partial leverage
regression plot 75, 219
Particle size and sampling 469
Particulate materials sampling 469
Partielle:
Information 879
Korrelationen 210, 658
Partitioned matrix, e. g. 54, 1207
Partitioning:
log-linear models, e. g. 1110
of chi-square 103, 263, 352, 793,
836, 1100, 1110, 1153, 1334
Partition of quadratic forms, e. g. 538
see also quadratic forms
Partnerwahl und Partnerbeziehung
1197
Part wearout and system deterioration, e. g. 32
Pascal-Verteilung 504, 631
Path:
analysis, e. g. 425, 522, 538, 628,
634, 662
diagrams illustrating possible
causal connections among
variables, e. g. 352
models, multivariate, e. g. 1277
sets: see minimal cut and path sets
Patient selection for a clinical trial,
e. g. 160

Pattern recognition 490, 612, 717,
964, 1083, 1085
applications of … in radar 717
in chemistry 377, 717
techniques for remote sensing
applications 717
Patterns of similarity: see cluster
analysis
PCBs, polychlorinated biphenyls and
other hazardous chemicals, e. g.
198
Peaks or wave forms, resolution of
overlapping 377
Pearl pregnancy rate, e. g. 1428
Pearson system of distributions, e. g.
541
Peeling bivariate data 47
Percentile equivalents of effect sizes,
e. g. 414
Percentiles:
and percentile ranks e. g. 739
and their standard errors, e. g. 739
Percent-percent plot, P-P plot: see
there
Percolation, generalized, for data
analysis and pattern recognition
612
Period:
age and cohort models and effects
in demography 851
analysis of population change 282,
1428
Periodenziffer in der Querschnittana-
lyse, z. B. 505
Periodogramm-Analyse 185, 186,
859, 1184
Permutational symmetry in
correlated variables, testing 94
Permutationen:
und Kryptographie, siehe dort
zufällige, z. B. 921
Permutation:
invariant procedures, e. g. 1072
tests 322, 629, 1072, 1085
significance levels and
confidence intervals for 629
Personen-Auswahlverfahren nach

dem Namen bzw. nach dem
Geburtstag 1166 bzw. 1167
Personnel flows within and among
organizations 50
Perturbation theory, e. g. 749
Perzentile und ihr Standardfehler 739
Peto and Peto's:
generalized Wilcoxon two-sample
rank sum test, e. g. 770
logrank test 21, 694, 770, 893, 1291
Petri-Netze 410, 529, 1016, 1104, 1105
Informationssysteme im Kranken-
haus 86
zur Tumornachsorge 692
Pfadanalyse, z. B. 425, 522, 538, 628,
634, 662
Pflanzenwachstum: statistische
Methoden, z. B. 178
Pharmacological calculations with
computer programs 1286
Pharmacology: applied statistics in
88, 160, 1353
Pharmakokinetische Studien, z. B.
1353
Pharmazeutische Industrie: stati-
stische Methoden 88, 160, 1353
Phi correlation 210, 367, 739, 798,
836, 947
Physical constants, errors in 1435
Pictorial charts, e. g. 1188, 1189
Piecewise-linear functions as limiting
forms, curve fitting with 1082
Pitman correlation coefficient, e. g.
836
Placebo, e. g. 160, 457, 692
Place/transition nets, e. g. 1105
Plackett's coefficient of association:
using in contingency tables with
ordered categories 1357
Planar-rotation algorithms versus
Gauss-Jordan methods, e. g. 890
Planning:
of experiments 474, 477, 1076, 1257
of observational studies 21, 205
of surveys: see surveys
Planung:
komplexer innovativer Systeme,
z. B. 1033

und Darstellung dynamischer und
konkurrierender Abläufe anhand von Petri-Netzen 410,
529, 1104
und Statistik 192, 205
von Erhebungen: siehe dort
von Experimenten 474, 477, 1076,
1257
von Untersuchungen in der Soziologie, z. B. 117
Plazebo, z. B. 160, 692
Plot:
Andrews 342, 351, 840
binomial data 546
biplot 47, 435, 513, 755
box plot variations: see there
C_P 83, 249, 300, 351, 890, 1383
cumulative weighted spacings 1025
diagnostic 547
half-normal 249, 300, 1440
H-H plot to investigate the
assumption of proportional
hazard rates of two samples
363
inside-out 1359
Mallows' C_P 83, 249, 300, 351, 890,
1383
marginal dot diagrams and scatter
plot 634
normal, e. g. 300, 1383
of success rates against centers of
gravity 916
parameter space 363
Poissonnes 546
principal component plot and aug-
mented principal component
plot 442
probability: see probability plots
radial 441
rankit 472, 1383
spread-versus-level, e. g. 547
straightness: see transformations
to linearize data
transformation, for additivity,
equal spread, straightness,
or symmetry 547
window 23

Plots:
Andrews plots of multivariate data 342, 351, 634, 840
and tests for symmetry 290
assessing multivariate normality 342, 634
box plot variations 23, 502, 547, 866, 1024, 1319, 1320, 1339, 1359
Chernoff-type faces 370, 1359, 1366
coefficient 566
comparison or schematic, e.g. 871, 1320
component and component-plus-residual plots 249
contour, e.g. 224
diagnostic 249, 547
for higher-dimensional data 417, 1366
see also Chernoff-type faces and graphical methods: visual display of multivariate data
Plots for symmetry 290
Plots, half-normal 249, 300
Plots: hazard plotting 332, 437, 770, 950, 952
Plots, influence-enhanced scatter 1296
Plots in regression analysis, e.g. 34
normal:
see normal probability plots probability 286, 300, 821, 1153, 1238, 1383
of drug potencies against storage times at elevated temperatures 160
of paired counts 915
of pairs of rotated factors, e.g. 1284
of STARS (or stars) 351, 634, 1359
of Studentized residuals 249, 300, 548, 1264, 1383
of the principal components, e.g. 634
on binomial probability paper 915
oval ... of cyclic data 1319

partial:
leverage regression plots 75, 219
regression plots 1316
residual plots 219
P-P and Q-Q 363, 546, 643, 821, 952, 1266, 1359
probability 219, 343, 363, 546, 643, 760, 821, 952, 1238, 1266, 1359
stabilized 886
Q-Q 69, 70, 363, 634, 643, 755, 1359
quantile 69, 70, 363, 634, 643, 755, 1359
rankit 472, 1383
relationships among variables, e.g. 417
representing points in many dimensions by trees and castles 683
residual 75, 219, 249, 300, 351, 442, 547, 550, 643, 755, 1077, 1264, 1340, 1383
rotated factor, e.g. 1284
scatter 201, 219, 249, 363, 502, 755, 828, 916, 1122, 1284, 1296, 1320, 1359
schematic, e.g. 871, 1320
sunflower 201
time series, e.g. 871
transformation 547
Plotting:
see also graphical methods, graphische Methoden
and displaying multivariate data 69, 70, 342, 351, 370, 1317, 1359, 1366
growth data 566
hazard 332, 437, 770, 950, 952
in exploratory data analysis, e.g. 1339
multi-dimensional data 69, 70, 183, 342, 351, 370, 683, 1209, 1317, 1319, 1359, 1366
of harmonic curves 342, 351, 840
of means with uncertainty intervals 23, 1319
on various scales to select the appropriate variance stabilizing transformation 121, 1153

probability: see probability plots
ranked observations on probability
    paper and drawing lines 821
residuals 75, 219, 249, 300, 351,
    442, 547, 550, 755, 1077, 1264,
    1340, 1383
s or s^2 versus $\bar{x}$ 121, 1153
STARS 351, 634, 1359
two-way-fit 1319
Plot, window 23
PLUS analyses 547, 916, 1320
Point:
    correlation coefficient, e.g. 1065
    estimation, e.g. 48, 66, 774, 1028,
        1090
    processes, e.g. 233, 1118
    process models in reliability, e.g.
        32
Points: see central tendency of a pat-
    tern of
Poisson:
    -binomial distribution 298, 542,
        631, 632, 701, 1223
    binomial distribution: fitting the,
        e.g. 1223
    -binomial mixture distributions
        343
    -Binomial-Verteilung 298, 542, 631,
        632, 701, 1223
    distribution 99, 103, 472, 504, 542,
        631, 632, 658, 701, 1028, 1056
        generalizations 540, 542, 1394
        mutation frequencies, statistical
            significance and power
            1443
    distributions, generalized, simpli-
        fied techniques for estimating
        parameters of some 540
    distribution used in testing wheth-
        er mutation frequencies differ
        from control frequencies 1443
    means, ratio of two, confidence in-
        tervals for the 948
    model fitting 546
Poissonness plot 546
Poisson:
    parameters, improving the normal

approximation when
    constructing one-sided
    confidence intervals for 486
populations, inference on trends in
    several 58
predictor intervals 948
process, e.g. 53, 233, 760, 874,
    1136, 1304
    compound, e.g. 53, 64
Poisson rates, multiple comparisons
    for 1174
    ratios for single sampling plans 984
    regression:
        analysis 385
        models in estimating incidence
            rates and ratios 386
    variables, tolerance interval for 479
Poisson variates: the analysis of ra-
    tios and cross-product ratios 396
    -Verteilung 103, 504, 542, 631, 632,
        658, 701, 1028, 1056
        ausführliches Beispiel 280
        confidence interval for the ratio
            of two Poisson means and
            Poisson predictor intervals
            948
        confidence interval for the ratio
            of two Poisson parameters
            1254
        graphische Methoden 546
        Verallgemeinerungen 540, 542,
            1394
    -Verteilungen, der Vergleich meh-
        rerer 146, 1174, 1443
Polishing 547, 1320
Polya-Aeppli distribution, e.g. 298,
    542
Polya's enumeration formula, e.g.
    1315
Polychoric correlation, e.g. 427
Polynomial:
    distributions: multiple compari-
        sons and comparisons with a
        control 1141
    regression 83, 285, 661, 902, 955,
        1122, 1390
        extension of: see spline
            regression

Polytomous variables and incomplete contingency tables 1334

Pooled cross-sectional and time series data analysis 274

Popper's demarcation between science and pseudoscience, e.g. 291, 1401

Population:
attributable risk 137, 151, 338, 367, 1363, 1364, 1429
change 166, 340, 455, 601, 1232, 1281, 1428
for Australia 1911 to 1966, e.g. 1281
closed, estimating the size 103, 966, 1000, 1064, 1065, 1208, 1210, 1391
dynamics 667, 966, 1281
exploitation of a plaice population 1065
growth, e.g. 601, 1065, 1281
healthy, reference values in laboratory medicine for a 441
modeling of a natural, e.g. 1065
sampled population and target population, differences between 205

Populations:
ageing, in Britain and USA 818
biological, sampling 223
extinction and formation of, e.g. 1065
finite, sample surveys of, the development of 35

Population-size estimation 103, 966, 1000, 1064, 1065, 1208, 1210, 1391

Populations:
multivariate, structuring 1078
ranking and selection of 74, 307, 406, 407, 465, 466
selection of 405–407, 465, 466

Population:
stable population theory 601, 1232
standard or norms, e.g. 422
statistics 340, 591, 601, 667, 676, 750, 818, 851, 1148, 1255, 1281, 1404, 1428

of Denmark, e.g. 601
theory, stable, extensions of, e.g. 601, 750
what … does a sample represent? 35

Portfolio analysis, stable 1078

Possibility theory and fuzzy sets 1367

Post-data model construction 1046

posteriori-Verteilung und Bayes-Statistik 684

Post-stratification 1237

Poststratification in clinical trials, e.g. 1214

Power:
analysis in game theory, e.g. 128
and sample size calculation using noncentral F approximations 462
and the detection of outliers 607
curves for the analysis of means 946
functions, e.g. 285
and their properties, e.g. 1090
noncentral distributions and power calculations 462, 489, 514, 633, 1005, 1032
of a test 286, 461, 462, 525, 658, 1072, 1323
of chi-square tests for 2×2 contingency tables with small expected frequencies 1002
of procedures for comparing two groups on p variables 1089
of univariate pairwise multiple comparison procedures 1088
paired versus two-sample design, treatments with dichotomous outcome 1354
spectrum estimation 718
the formalization of the notion of political power in the context of two person negotiations 128
transformations, e.g. 547, 716, 916, 1052, 1320, 1390

P-P and Q-Q plots for comparing two

distribution functions: see P-P plots and Q-Q plots

P-P plots 219, 343, 363, 546, 643, 821, 952, 1238, 1266, 1359

Prädiktiver Wert und Effizienz in medizinischen Diagnosen 367, 390, 986, 1355

Prämienkalkulation in der Versicherung 711

Prärandomisierung, prärandomisierte Versuchspläne 620

Prävalenz, prevalence, e. g. 390, 682

Prävalenztabellen 390

Pre- and post-treatment data analysis, e. g. 21

Precision:
    and accuracy:
        in clinical chemistry 172, 289, 441, 1387
        in measurement 279, 733, 1081
        see also calibration
    measurement:
        and calibration, e. g. 733
        and tolerance regions for relative errors 624

Predictability of two members of a system, asymmetry in 1365

Predicting a single score from many others: see multiple regression analysis

Prediction:
    analysis of covariance and standardization as instances of 751
    and entropy 35
    by use of a logit model, e. g. 751
    influence, e. g. 219
    intervals:
        a survey of … and their applications 481
        for screening 1008
        on the distribution of the Studentized maximum of equally correlated normal random variables 467
        sample sizes for 876
    logic for nominal data, e. g. 1110

Predictions from multivariate regression: confidence and prediction ellipses, e. g. 634

Predictive value in medical diagnosis 213, 367, 390, 986, 1355

Predictor variables, highly correlated 555, 663, 843, 956, 1350

Predictors, principal components of, e. g. 1015

Preference and similarity analysis 612

Pregibon's diagnostics for generalized linear models illustrated with an example on logistic regression 219

Preisstatistik 20, 268, 671, 1106, 1171

Premium calculation principles in insurance 431

Presentation and interpretation of statistical results in medical research, e. g. 888

Presenting a statistical paper 328, 379, 1263

PRESS, prediction sum of squares, e. g. 15, 300

Prevalence, e. g. 213, 390, 682

Prices, wages, manpower, …, statistics 20, 671, 1106

Pricing of insurance products 431

Principal:
    component:
        plot and augmented principal component plot 442
        regression, e. g. 300
    components analysis 8, 22, 54, 83, 370, 501, 513, 604, 634, 661, 767, 800, 814, 831, 835, 840, 855, 858, 922, 1078, 1085, 1284, 1436
        an application of two generalizations of non-linear 612
        for highly skewed distributed variables, correspondence analysis versus 337
        generalizations of non-linear 612

Principal:
   component:
      of qualitative data 47, 164, 448,
         535, 596, 800, 835, 840, 965
      versus/and factor analysis 604,
         1284
      versus correspondence analysis
         337
      components, robust: assessment of
         multivariate normality 853
      co-ordinates analysis 342, 435
Prior distribution in Bayesian
      reliability analysis 847
priori-Verteilung und Bayes-Statistik
      684
PRISM 409
Privater Verbrauch: Ermittlung und
      Fehlerabschätzung 1102
Privathaushalte:
   Daten 591, 1079, 1255
   Vorausschätzung der 803
Probabilistic developments, new 35
Probabilities:
   calculating by using recurrence
      relations 904
   calibration of 640
   combining, e.g. 414, 1132, 1153,
      1270, 1362
   compound:
      correlated, bounds for 1377
      evaluation of … in sequential
         choice 640
   compounding … from indepen-
      dent significance tests 414,
      1132, 1153, 1270, 1362
   rare event, assessing risks through
      the determination of 1160
Probability:
   and tennis 407, 1062
   density estimation 47, 401, 658,
      1288, 1386
   models 19, 118, 304, 454, 643, 768,
      816, 823, 1093, 1153
   of an event as a function of dicho-
      tomous or polytomous data,
      estimation of the 1360
   of winning a game, set or match of
      classical and tie-breaker
      tennis 1062

on the history of: see history of sta-
      tistics
plots 219, 343, 363, 546, 643, 760,
      770, 806, 821, 952, 1052, 1238,
      1266, 1359, 1390
   and plotting positions, e.g. 808
   interpretation of, e.g. 1266
   stabilized 886
sampling 621, 729, 1274
statements, falsifying, e.g. 291
subjective, e.g. 48, 498
   see also subjective probability
the calculus of 19, 118, 304, 454,
      498, 643, 768, 816, 823, 1093,
      1153
theories of, e.g. 498
types of, e.g. 48, 498
Probitanalyse von Prozentzahlen 13,
      414, 531, 799, 864, 1332, 1374
Probit analysis, probit models 13,
      414, 531, 864
Problembeschreibung und Modell-
      bildung, z.B. 1150
Problemlösen und planen, z.B. 192,
      423, 1033, 1194
Problemlösungsbaum, z.B. 1033
Problems, anomalous 754
Process:
   control, e.g. 781
   variation studies: graphical
      analysis 1239
Procrustes:
   analysis 435
   rotation, e.g. 840
Production and productivity (statis-
      tics) 1106
Productivity measurement: the
      measurement of total factor
      productivity 614
Product:
   life:
      analysis: see survival data
         analysis
      distributions, e.g. 99, 420, 952
   limit estimator: see Kaplan-
      Meier's estimator
   -moment correlation coefficient:
      see correlation coefficient

Produktmoment-Korrelationskoeffizient: siehe Korrelationskoeffizient

Produzierendes Gewerbe: Statistik 20, 268, 1255

Profile:
  analysis, e.g. 44, 634
  data in multivariate analysis, e.g. 513

Prognoserechnung, Prognostik 150, 199, 261, 526, 557, 784, 785, 803, 880, 881, 1184, 1203

Prognose- und Entscheidungsfindung in der Medizin 620

Prognosis and diagnosis with tumor markers 2, 986, 1355

Prognostic:
  evaluation of refence values 441
  factors:
    identification of, e.g. 770
    in sequential clinical trials 1393
  tests in medicine 250, 986

Prognostische Tests in der Medizin 250, 986

Programmierung, lineare, nichtlineare und dynamische, z.B. 333, 490, 536, 860, 1158, 1308

Programming:
  dynamic, geometric, integer, linear, quadratic and separable 1308
  integer, e.g. 536, 1308
  linear, e.g. 490, 536, 1158, 1308
  nonlinear, e.g. 536, 860
    linearly and nonlinearly constrained problems, e.g. 860
    sequential, unconstrained minimization techniques, e.g. 860

Programmpakete: siehe Programmsysteme

Programmsysteme 83, 93, 114, 206, 252, 285, 376, 404, 478, 639, 661, 741, 786, 864, 891, 897, 918, 926, 1001, 1121, 1122, 1129, 1213, 1284, 1327, 1345, 1412
  siehe auch BMDP, GLIM, IDA, SAS, SPSS und Expertensysteme

Ergänzung durch wichtige statistische Algorithmen, z.B. 453

Programs for variance estimation 1419

Progress in science, scientific method 123, 182, 220, 229, 291, 303, 423, 525, 754, 882, 1024, 1066, 1318, 1321–1323, 1361, 1404, 1411

Progressive diseases, models of, e.g. 419

Projection pursuit methods, e.g. 755

Projektmanagement-Methoden, z.B. 1033

PROLOG, e.g. 26, 1163

Propensity scores 1131

Proportional:
  hazard estimation 11, 144, 234, 649, 651, 770, 862, 1060
  hazards:
    assumption, checks on the 363, 649
    model: survival data analysis, e.g. 234
    regression models 173, 228, 234, 285, 642, 651, 862, 893, 1182
  linear models 241
  odds 862
  -reduction-in-error measures, e.g. 1110

Proportions:
  and the logit scale: see logit models
  binomial, confidence intervals for the ratio of two 697
  chi-square tests of homogeneity of proportions for qualitative variables in independent samples, e.g. 836, 1153
  comparing two:
    see also fourfold table analysis
    finding p_2 when p_1, n, alpha and beta are specified 1337
  confidence limits for 105, 984, 1426
  ratio of two, confidence intervals for the 697
  standardization of a logit-linear additive table of, e.g. 805
  tables of tests and confidence limits 984

Proportions:
two and their odds ratios, a table of exact confidence limits for differences and ratios of 1299
with more than two predictors, analysis of 374
Prospective study: see cohort study
Prozentzahlen:
EDV-Auswertung, Logit- und Probitanalyse 799
und Likelihood-Schätzungen 799
Prozeßrechner, z. B. 1193, 1312
Prüfverteilungen, (t, χ^2, F u. a. m.) 504, 631, 658
Pseudo-Bayes estimates of cell probabilities, e. g. 103
P-STAT: see e. g. factor analysis
Psychiatric statistics 35
Psychometric applications of log-linear models 355
Public policy and statistics 913
Punktbiseriale Korrelation bei Bindungen 154
Punktbiserialer Spearman-Test 154
P value, e. g. 408, 1072
P values, ensemble-adjusted 1133
P-Wert, z. B. 408, 1072

Q-Q-plots 69, 70, 363, 634, 643, 755, 1359
Q-Q-plotting, multiple 755
Q-Test:
nach Cochran 709, 766, 793, 836
lineare Kontraste 787
Quadratic forms, e. g. 54, 443, 538, 634, 716, 732, 789, 1207
Quadratische Kontingenztafeln, siehe square contingency tables
Qualitative data:
analysis, e. g. 889
quantitative analysis of, e. g. 97, 660, 1137, 1436
see e. g. contingency table analysis and correspondence analysis
Qualitative Daten, multivariate Analysen: Aufdeckung von Schein-

zusammenhängen durch den GSK-Ansatz: siehe dort
Qualitative variables, tests of intra-class dependence for 60
Quality control 160, 460, 700, 781, 850, 1008, 1181, 1329, 1389
and reliability 722
modern developments in 781
Quality of data 36, 160, 197, 932, 1074, 1076, 1185, 1243, 1424
Quantal:
assays 160, 359, 360, 531 , 582, 1095, 1332
response data: life data analysis for units inspected once for failure, e. g. 951
responses in biology 173, 359, 360, 531, 582
Quantile:
bootstrap 1025
-box plots 547, 1024, 1025
functions in statistical modeling, e. g. 547, 1024, 1025
plots 363, 643, 1359
for assessing symmetry following Wilk and Gnanadesikan, Tukey or Doksum 363
-quantile plot, Q-Q plot: see there
simulation and quantile bootstrap 1025
values for data summary, e. g. 547, 1025, 1339
Quantile: verteilungsfreie Vertrauensgrenzen für Quantile stetiger Verteilungen, z. B. 1426
Quantitative methods in social sciences, e. g. 650
Quasi-:
Experimente 117, 220, 1404
independence in two-way tables, e. g. 472, 1110
likelihood:
functions 863
models, e. g. 864
Quellen statistischer Daten: international und USA 1372

Querschnittstudien, z. B. 274
Query:
　optimization in data bank systems
　　1329
　processing in data base systems 61,
　　670, 1330
Question framing and response
　consistency 558
Questionnaire design 36, 125, 167,
　609, 1137, 1182, 1200, 1276
　and analysis, e. g. 206
Questions:
　answering sensitive or highly
　　personal questions under
　　retained privacy: see
　　randomized response tech-
　　niques
　influences on survey responses
　　646, 1200, 1424
　in surveys 36, 125, 167, 564, 609,
　　646, 1137, 1200, 1276, 1424
　response effects to behavior and
　　attitude questions 1424
　sensitive: see randomized response
　　techniques
Queueing:
　models 53, 339, 347, 454, 536, 695,
　　760, 874, 1136
　　statistical inference for, e. g. 53,
　　　339, 536, 1136
　networks and reversability 1136
Queues, general, approximate
　waiting times in more 1014
Quota sampling 621, 729, 820, 1061,
　1137, 1201, 1237, 1261, 1274,
　1431, 1433

Radial plot of multivariate data 441
Radioimmunoassay 360
Random:
　digit dialing in selecting a popula-
　　tion-based control group 497
　effects and mixed models in analy-
　　sis of variance 31, 264, 896
　fields 35

Randomisierung 121, 661, 690, 1058,
　1151, 1152, 1153
Randomisierungstests 322, 629, 1072
　für eine Person 322
Randomization 21, 121, 367, 477, 538,
　661, 690, 1058, 1151, 1153
　device: application of the random-
　　ized response in surveys, e. g.
　　1260
　in clinical trials, e. g. 770
　methods/tests e. g. 322, 326, 629,
　　1072, 1085
　procedures, e. g. 322, 326, 629,
　　1072, 1085
　restricted, versus blocking 40
　tests, one-sample and two-sample,
　　e. g. 1072
Randomized:
　allocation of treatments in sequen-
　　tial medical trials 55
　block design, e. g. 83, 121, 674, 681,
　　955
　　tables of the distributions of
　　　rank statistics for treatment
　　　versus control 980
　design, completely, e. g. 83, 674,
　　955
　response:
　　methods for obtaining sensitive
　　　quantitative data,
　　　scrambled 329
　　models for estimating the distri-
　　　bution function of a quanti-
　　　tative character 311
　　techniques, e. g. 311, 329, 820,
　　　1137, 1260
Randomness against:
　autocorrelation, tests of 688
　trend or serial correlations, tests of
　　719
Randomness:
　and time series 312
　how to achieve, in sampling, e. g.
　　1395
　of a sequence of dichotomous data
　　or measured data, testing the
　　1152, 1153

Random:
    number generation and testing 129,
        130, 308, 661, 760, 908, 1085,
        1097, 1117, 1143, 1190
    numbers: generation and testing
        with TESTRAND 308
    number generators for a pocket
        calculator 1338
    numbers, normal, generation of,
        e.g. 421, 661, 1035
    processes: applications and
        worked examples, e.g. 454
    sampling 98, 621, 658, 727, 729,
        1061, 1169, 1201, 1261, 1271,
        1274, 1395, 1431, 1433
    sequence: detecting a scale shift
        1287
    variables, e.g. 454, 643, 768, 769
    variate generation, e.g. 129, 760,
        1085, 1117, 1143
    variates generation from the com-
        mon continuous and discrete
        distributions and from multi-
        variate distributions, e.g. 661
    vectors, basis properties of, e.g.
        840
    walks, e.g. 454, 643, 1136
        and martingales 454, 1136
Rangkorrelation 374, 433, 434, 439,
        440, 597, 766, 775, 793, 836
    multiple und partielle, z.B. 775
    nach Spearman 317, 439, 597, 766,
        836, 1152, 1153
Rangordnungen, Dominanz- und
        Ähnlichkeits-Paarvergleiche,
        z.B. 117
Rangordnungsstatistik 214, 253, 495,
        658, 793, 836, 1209
Rangreihen: Maße für die Überein-
        stimmung 433, 434, 790, 1018
Rangsummentest: siehe verteilungs-
        freie Methoden
Rangtransformation 216, 217
Rangvarianzanalyse, mehrfaktorielle,
        bei ungleichen Zellbesetzungen
        578, 579
Rank analysis of covariance 585, 764

Rank correlation (methods) 374, 433,
        434, 439, 440, 597, 766, 775, 793,
        836, 1153
Rank correlation:
    agreement between rankings 433,
        434
    Kendall's tau 285, 440, 766, 793,
        836
    multiple, e.g. 775
    nomograms 439, 440
    partial, e.g. 775
    Spearman's rho 285, 439, 597, 766,
        836, 1153
Ranking:
    after alignment procedures 543
    and selection of populations 74,
        307, 406, 407, 465, 466
Rankings:
    concordance of 433, 434, 790, 1018
    see also agreement of, rank corre-
        lation, and Friedman test
Rankit:
    plot 472
    plots or normal plots, e.g. 1383
Rank:
    methods for various statistical
        analyses, e.g. 214–217, 253,
        495, 719, 793, 836, 1072
    statistics and limit theorems 719
    test for correlated samples of
        different sizes 94
    tests for complete block designs,
        e.g. 509
    tests in linear models 719
    transformation 214–217, 620
Rare:
    diseases: a comparison of sample
        size determination methods in
        the two group trial 780
    event probabilities and risks 1160
    events, relatively rare events: four-
        fold tables 1002
Rates, adjusted:
    directly and indirectly 194, 380
    with their standard errors 194
Rates:
    age-adjustment of, e.g. 602

and proportions 367
failure 21, 99, 420, 544, 770, 951, 1226
Poisson, multiple comparisons for 1174
standard errors of infant mortality rate, age-adjusted death rate, age-specific death rate, survival probability, the expectation of life, and the difference of two estimates of probabilities 194
standardized 21, 103, 193, 357, 367, 677, 678
standard or standardized: see also Standardisierung
summary, for reporting health data 380
Rating:
experience, e.g. 575
-Skalen, z.B. 117
Ratio:
estimator, estimation of variance of the: an empirical study 123
of the geometric mean to the arithmetic mean as an unorthodox measure of relative dispersion in a distribution 613
of two binomial proportions, confidence intervals for 697
of two Poisson parameters, calculating a confidence interval for the 1254
Ratios and differences of two proportions and their odds ratios, a table of exact confidence limits for 1299
Ratten: Lernversuche 1353
Raumordnung und Infrastruktur 161
Rayleigh:
distribution, e.g. 99, 541
test, e.g. 840
$r \cdot c$ table, see contingency table
Reaktionszeiten, z.B. 184
Realzeitsysteme: Zuverlässigkeit und Fehlermanagement 1282
Rechnernetzwerk 61

Rechtspflege 1079, 1255
Reciprocal averaging, e.g. 965
Recurrence relations, e.g. 1315
for calculating probabilities 904
Recurrence:
risk to disease, e.g. 538
times, the distribution of, e.g. 420
Recursive least squares, e.g. 271
Redundancy:
and standby redundancy, e.g. 99
to achieve fault tolerance, use of, e.g. 46
Reed-Frost theory of epidemics: classic model and a modified model with two intermixing populations 171
Re-expression of data 502, 831, 871, 916, 1320, 1323
Reference:
limits determination, parametric and nonparametric 0.90 confidence limits of the 0.025 and the 0.975 fractiles, e.g. 441
values in laboratory medicine 441
Regenerative method for simulation analysis, e.g. 1143
Regionalstruktur der Bevölkerung 37, 340
Regression:
adjustments 21, 205
adjustment versus nearest available matching 205
Regression analysis 24, 83, 93, 114, 300, 349, 468, 549, 551, 798, 824, 831, 895, 916, 955, 956, 1001, 1085, 1093, 1195, 1238, 1263, 1350, 1383, 1390, 1420
application of the singular value decomposition in 417, 831
calibration 148, 318, 319, 395, 577, 610, 685, 733, 815, 1176, 1280, 1295, 1324
choice of subsets in, e.g. 300, 549, 550, 890
collinearity 75, 551, 831, 1316, 1340, 1383
and variable selection 1383

Regression:
  comparing:
    more than two regression lines
      1153
    nonparallel regression lines
      1128
    response curves 44, 178, 703,
      709, 772, 1041
    two dependent regression lines
      1236
    two regression lines 1251
  correlated errors in: see autocorre-
    lation
  C_p-statistic 4, 83, 249, 300, 351, 890,
    1383, 1437
  diagnostic, graphical methods of
    34
  dummy variables 300, 564, 681,
    798, 955, 1383, 1437
  heterogeneity of variance:
    symptoms and remedies 1383
  heteroscedasticity 955, 956, 1350
  influential data points in 75, 219,
    300, 442, 755, 956, 990, 1264,
    1316, 1340, 1383
  linear equality restrictions 1115
  making predictions, e.g. 1383
  multiple 8, 83, 93, 114, 206, 210,
    285, 288, 295, 300, 349, 482,
    663, 780, 798, 824, 831, 835,
    890, 899, 902, 955, 956, 1001,
    1238, 1284, 1383, 1390, 1420
  nonleast squares estimation: see
    ridge regression and robust
    regression
  nonlinear: see nonlinear regression
    models
  nonlinearity: symptoms and
    remedies 1383
  of cross-classification 1110
  of survival data: see survival data
    analysis
  ordinal data 18, 862
  outliers and influential data points
    75, 219, 1264, 1340, 1383
  Poisson 385, 386
  recent advances in 1350

  re-expression of the independent
    variable or of the dependent
    variable for linearizing the
    data 83, 300, 547, 1001, 1153,
    1383
  residuals: see there and residual
    analysis
  ridge regression 555, 663, 843, 956,
    1350
  selection of variables in nested
    data, e.g. 249
  strong interrelationships between
    the explanatory variables in:
    see multicollinearity
  two dependent regression lines,
    testing concurrence and
    parallelism of 1236
  two-phase regression 1212
  use of the singular value
    decomposition in 417, 831
  with a categorized explanatory
    variable 650
Regression:
  and association procedures, non-
    parametric graphical 363
  and autocorrelation, e.g. 121, 650,
    955
  and/or correlation 1369
  and econometric methods, e.g. 826
  and experimental design, the
    interrelationship between,
    e.g. 83
  and exponential smoothing
    methods to forecast non-
    seasonal and seasonal time
    series, e.g. 4
  and ordered categorical variables,
    e.g. 18, 862
  and residuals, e.g. 219, 249, 300, 547
    see also residual analysis
  approximate regression models
    and splines 300, 341
  bivariate 798
  Brown-Mood median regression
    line 363, 547
  coefficients, standardized, or beta
    weights, e.g. 210, 1284

constrained: see weighted least
squares procedures
curvilinear, e.g. 732, 1238
  see also nonlinear regression
diagnostics 75, 219, 442, 550, 755,
    1043, 1077, 1264, 1340, 1383
durch den Nullpunkt 174, 476
empirische 300, 1041
for dichotomous dependent
    variables: see logit models
    and probit analysis
homogeneity under variance
    heterogeneity, testing for 301
incomplete data, e.g. 1383
influential data points and outliers
    990, 1264, 1383
influential subsets in k-clustering
    as a detection tool for 442
intercept: see there
inverse: see calibration
kategoriale 346
least absolute value estimation, an
    annotated bibliography 275
leverage and influence in 75, 174,
    219, 442, 547, 550, 1264, 1316,
    1340
line:
  resistant 547, 1339
  with errors in both variables 147,
    300, 641
  with no intercept term, fitting a
    476
linear 15, 300, 549–551, 658, 832,
    902, 956, 1437
  determination of limits for a ...
    or calibration curve 577,
    1280
  determining lack of fit or func-
    tional misspecification 935
  developments in methodology
    300, 550
lineare:
  mit Prozentzahlen, z.B. 799
  Weiterentwicklungen 300, 550
linear:
  equality restrictions in 1115
  errors in the predictors and in
    the response 147, 300, 641

lack of fit or functional
    misspecification associated
    with the deterministic
    portion of a proposed
    linear regression model 935
least absolute value estimation
    in 275
lower limits 1280
nonparametric one-, two-, and
    k-sample tests for slope 436
nonparametric tests for intercept
    819
prediction 1383
robust, e.g. 147, 273, 300, 547,
    580, 587, 641, 929
selection of variables, e.g. 300,
    549, 550, 890
testing for lack of fit in 935
testing intercept parameter
    hypotheses by simple
    nonparametric methods
    819
with both variables in error 147,
    300, 547, 580, 587, 641, 832
lines, comparison:
  of nonparallel 1128
  of three over a finite interval
    530, 1153
  of two dependent 1236
  of two over a finite interval 1153,
    1251
lines:
  dependent, testing concurrence
    and parallelism of two 1236
  nonparallel, comparing 1128
  testing the equality of more than
    two 1153
logistic 8, 18, 135, 137, 173, 219,
    227, 288, 682, 813, 1182, 1314
logit, see logitlineare Modelle and
    Logitmodelle
maximum likelihood, program
    using the generalized F
    distribution 560
methods for forecasting non-
    seasonal and seasonal series,
    e.g. 4

Regression:
  methods, recent advances in 1350
  minimum sum of absolute error regression, a survey 929
  model: detecting departures 75, 219, 916, 1340
  model for bias in differences: mean matching 205
  models, e.g. 34, 300, 1420
    and life tables, e.g. 228
    and non-proportional hazards in the analysis of breast cancer survival 437
    and one or more switching points, e.g. 1087
    and survival prediction 994
    assessing, graphical methods, e.g. 183
    for experimental data, e.g. 15
    for grouped populations in cross-section surveys 1050
    for ordinal data 18, 862
    least absolute value estimation in 275
    leverage in 75, 174, 219, 547, 550, 1264, 1316, 1340
    nonlinear, parameter estimation in 716
    proportional hazards 173, 228, 234, 651, 862, 893, 1182
    proportional hazards, and the analysis of censored survival data 234, 649, 651, 770, 862, 1060
    switching a modification to tests in relation to the problem of 1087
    tests in relation to the problem of switching 1087
    with no intercept term, fitting 476
  multiple 8, 83, 93, 114, 206, 210, 285, 288, 295, 300, 349, 482, 663, 780, 798, 824, 831, 835, 890, 899, 902, 955, 956, 1001, 1238, 1284, 1383, 1390, 1420, 1437
    best subset regression procedures, e.g. 300, 549, 550, 890
    dealing with missing values 295
    mit Prozentzahlen, z. B. 799
    selection of subsets of regression variables, e.g. 300, 549, 550, 890
    stepwise and setwise as model building procedures, e.g. 1284
    under linear restrictions, e.g. 661, 1115
  multivariate, assessing dimensionality in 716
  multivariate, predictions from: confidence and prediction ellipses, e.g. 634
  nichtlineare 8, 249, 300, 587, 651, 824, 956, 1082, 1238
  nichtparametrische 590
  nonconstant variance … techniques for calibration-curve based analysis 395
  nonhomogeneous error variances, estimation and hypothesis testing 259
  nonleast squares estimation: see ridge regression and robust regression
  nonlinear 8, 85, 249, 300, 587, 651, 661, 732, 824, 956, 1082, 1098, 1238
    robust, e.g. 587, 661
    unconstrained optimization and, e.g. 587, 661
  nonparametric 201, 590
  ordered categorical variables 18
  ordinal data 18, 862
  orthogonal polynomial 300, 661, 732, 930, 931
    for unequal spacing and frequencies 930
  outlier tests 75, 219, 1264, 1340, 1383
  polynomial 83, 285, 681, 902, 955, 1122, 1390
    extension of: see spline regression

principal components regression 831

program for microcomputers, e. g. 15

programs, see e. g. BMDP, SAS, SPSS

proportional hazards models and the analysis of censored survival data 234, 649, 651, 770, 862, 1060

random coefficient regression 274

residuals, examination of 219, 235, 249, 300, 547, 651, 761, 765, 915, 916, 955, 956, 1316, 1340, 1383, 1406

resistant, e. g. 547, 916, 1339

Ridge-Regression 31, 300, 555, 663, 843, 956, 1015, 1350, 1385

robust 147, 219, 273, 300, 417, 547, 550, 551, 580, 587, 641, 765, 829, 929, 1339

smoothing scatterplots and graphical methods, e. g. 755

tests of fit for normality, e. g. 808

to the mean, empirical Bayes estimator anticipates 175

transformations 83, 300, 547, 1001, 1383

two-phase ... and change point problem 1212

Regressionsanalyse 83, 93, 114, 285, 300, 349, 549, 551, 798, 831, 895, 916, 955, 956, 1001, 1093, 1195, 1238, 1263, 1350, 1383, 1390, 1420

    Anpassung einer Ausgleichsfunktion an beobachtete Werte 249, 891, 1041, 1376

    Kalibrierung, siehe dort

    lineare Regression 300, 549–551, 658, 902, 1437

    lineare, beide Variablen fehlerbehaftet 147, 300, 641

    mit Prozentzahlen 799

    multiple 8, 210, 300, 349, 482, 663, 798, 831, 899, 902, 955, 1383, 1390, 1420

nichthomogene Varianzen 259

ordinale Daten 18, 862

orthogonale Polynome 300, 732, 930, 931

Ridge Regression 555, 663, 843, 956, 1350

    und Pfadkoeffizienten-Analyse 1384

    Vergleich dreier Regressionsgeraden 530, 1153

    Vergleich empirischer Verlaufskurven 709, 772, 1041

    Vergleich zweier abhängiger Regressionsgeraden 1236

    Vergleich zweier Regressionsgeraden 1153, 1251

Regressionsebene und Regressionshyperebene 300, 349

Regression:

    segmented straight lines 300, 1082

    stepwise 300, 549, 550, 1383

Regressionshomogenität bei Varianzheterogenität 301

Regressionsmodelle für Ordinaldaten 18, 862

Regression: splines 300, 341

Regressionsrechnung 83, 93, 114, 300, 349, 549, 551, 658, 798, 831, 835, 895, 916, 955, 956, 1001, 1093, 1195, 1238, 1263, 1350, 1383, 1420

Regression:

    subsets in regression analysis, selection of 890

    systematic location of unusual data points 75

    tests of fit for normality 806

    three-group resistant line, e. g. 547

    three independent variables 249

    through the origin and leverage 174

    transformations in regression models 83, 300, 547, 1001, 1153, 1383

    trees and classification 132

    two-phase 1212

    variables, selection of subsets of, e. g. 300, 549, 550, 890, 1437

Regression:
 Vergleich dreier Regressionsgeraden 530, 1153
 with indicator variables, dummy variables or binary variables 300, 564, 798, 955, 1383
 with nested data 249
Reinsurance as a cooperative game 128
Relational data base systems 104, 576, 1330, 1349
Relationales Datenbankmodell 104, 576, 1330, 1349
Relational query languages 1330
Relationship:
 assessing the amount of relationship between a set of independent and a set of dependent variables: see canonical correlation analysis
 functional and structural, e.g. 658
Relationships:
 and causal relationship, e.g. 1153, 1437
 between variables 274, 502, 547, 658, 916, 1319, 1320, 1323
 complex, expressing … in two dimensions 47
Relative change and its measurement 1310
Relative efficiency, asymptotic, e.g. 836, 1072, 1090
Relative risk:
 (relatives Risiko) 6, 21, 80, 86, 134, 135, 137, 194, 338, 367, 561, 569, 574, 682, 697, 791, 1182, 1292, 1429
 and weak associations between an exposure and a disease 1429
 functions, general, for case-control studies 139
 index 602
 statistical tests and interval estimation 80
 summarizing, Poisson regression models for 386
 summary estimator of stratum-specific relative risks 682, 1292
 unconditional tests of hypotheses concerning the 80
Relative risks: test for heterogeneity of the stratum-specific relative risks 1292
Relatives Risiko: siehe relative risk
Relevanzbaum, z. B. 1033
Reliability analysis, e.g. 99, 420, 524, 1308
 see also: reliability of engineering systems
 Bayesian, e.g. 847, 1309
 optimal inspection and maintenance schedules for systems subject to failure 1216, 1217
Reliability:
 and availability modelling, e.g. 999
 and maintainability in the acquisitions process, e.g. 265
 and validity 739, 1230
 Bayesian reliability analysis 847, 1309
 distribution: a … with increasing, decreasing, constant and bathtub-shaped failure rates 544
 engineering:
  and risk assessment 378, 524
  data banks: see there
 evaluation with frequency and duration techniques, e.g. 99
 functions, e.g. 99
 growth, e.g. 32
  nonparametric estimates for, e.g. 265
 in the acquisitions process 265
 in the acquisitions process and recent work related to renewal theory, e.g. 265
 life testing, e.g. 265
 nonparametric concepts and methods in 719
 of computer software and computing networks 97
 of engineering systems 32, 46, 97, 99, 265, 420, 475, 524, 536, 544, 552, 691, 699, 762, 847,

96

883, 948, 949, 952, 974, 999, 1130, 1165, 1168, 1216, 1217, 1267, 1304, 1308, 1309, 1327
of repairable systems, e.g. 32, 99
of scientific inference 182, 229, 269, 291, 882, 1024, 1066, 1321, 1322
statistical methods in, e.g. 762
system(s), optimization of 999, 1308, 1327
theory, stochastic models in, for engineers e.g. 999
Relierter Likelihoodquotienten-Test 1393
Renewal process 32, 233, 420, 454, 695, 874, 1136, 1304
the renewal equation, limit theorems, excess life, and applications 454
Renewal theory:
and MTBF, e.g. 420
and reliability in the acquisitions process, e.g. 265
Monte Carlo simulations, e.g. 420
Repairable system reliability versus nonrepairable system reliability, e.g. 32
Repairable systems:
failure data, e.g. 32, 99, 420
reliability 32, 99, 265, 378, 1304, 1327
Repair and installation process, two stage, e.g. 99
Repair times, reliability models including, e.g. 32
Reparaturzeiten 32, 99, 552, 1168
Repeated chi-square testing 957, 1152, 1153
Repeated measurements:
design and analysis 190, 573, 690, 716, 1408
design: confidence interval for the largest mean 190
parametric and non-parametric analysis for 690
Repeated measures:
(models, design and analysis) 31,

44, 206, 322, 516, 634, 836, 856, 895, 896, 901, 972, 975, 1017, 1368
analyses, univariate, serial correlation in 516
and matched samples 836
MANOVA method for analyzing … designs 975
versus split plot designs 896, 901
Replacement: problems and modeling, e.g. 999
Report:
quality assurance and proofreading 160
writing a scientific 328, 379, 1263
Representative samples and representative sampling 729
Resampling plans: see jackknife techniques and bootstrap methods
Research:
cumulating and integrating research findings across studies: see meta-analysis
designs for social scientists, e.g. 1246
results: defining, retrieving and assessing, comparing and combining in social research 1132
structure 182, 229, 303, 423, 754, 882, 1066, 1318, 1321–1323, 1361, 1411
Resemblance coefficients: see cluster analysis
Residual-Analyse:
siehe residual analysis und residuals
des Unabhängigkeitsmodells zweier kategorialer Variablen 472, 620
Residual analysis 219, 235, 249, 300, 547, 651, 761, 765, 871, 911, 915, 916, 955, 956, 1284, 1316, 1340, 1383, 1406
in two-way-tables 219
Residual:
displays, the robustness of 756

Residual:
    lifetime curves of strikes, the shape
        of expected 763
    plots 75, 219, 249, 300, 351, 442,
        547, 550, 643, 755, 1077, 1264,
        1340, 1383
Residuals:
    see also analysis of residuals
    adjusted, e.g. 547
    analyzing ... in calibration
        problems 990
    and regression, e.g. 219, 249, 300,
        547, 1284, 1437
    in calibration problems, analyzing
        990
    in contingency table analysis 472,
        620
    serial correlation 300
    standardized 547, 1316, 1340
        and adjusted, for contingency
            tables, e.g. 472, 1110
Residuals, Studentized 249, 300, 547,
    548
Residual sum of squares and selec-
    tion of subsets of regression
    variables, e.g. 890
Residuen und ihre Analyse 219, 235,
    249, 300, 547, 651, 761, 765, 915,
    916, 955, 956, 1284, 1316, 1340,
    1383, 1406
Resistant:
    analysis of two-way tables: see
        median polish
    lines 547, 916, 1339
    statistics, resistance in statistics
        502, 547, 1323
    regression lines for y versus x, e.g.
        547, 916, 1339
Resolution designs, e.g. 121
Response:
    consistency and question framing
        558
    curves:
        and growth curves 44, 178, 703,
            708, 772, 1041
        comparision of 703, 708, 772,
            1041

        estimation of a mean response
            curve 241, 1041
        nonparametric comparison of k
            independent samples 772
        computer program for the distri-
            bution-free analysis of 703
        the comparison of two indepen-
            dent or two matched
            samples of 708
        timing of subsequent measure-
            ments of 205
        with non-equidistant points of
            time, comparison of two in-
            dependent or two matched
            samples by an orthogonal
            polynomial 708
    effects in surveys, e.g. 1137
    caused by threatening and
        nonthreatening behavior
        questions 1424
    error reduction 167, 594, 992, 1275,
        1424
    probabilities: models for response
        and nonresponse errors in
        surveys 594
    quantal, in biology 173, 359, 360,
        531, 582
    surface applications: the theory of
        "alphabetic optimality" is un-
        duly limiting 120
    surface design 120, 121, 173, 251,
        556, 873, 912, 1257, 1272
        robustness to bad values, e.g. 120
    surface model diagnosis in two-
        way tables 127
R-estimators, e. 547, 719
Restrictions in multiple regression,
    e.g. 661, 1115
Results: see also research results, ...,
    in social research
Retrospective study: see case control
    study
Reversal paradox 884
REX, Regression Expert, a project 478
Ridge analysis 556
Ridge regression 31, 271, 300, 555,
    663, 843, 956, 1015, 1350, 1385

knowledge of the predictor space is necessary 1015

Ridit analysis 367, 611, 692

Risiken technischer Systeme: siehe Zuverlässigkeit technischer Systeme

Risiko-Abschätzungen 84, 137, 225, 243, 332, 361, 375, 418, 524, 528, 561, 569, 638, 682, 864, 903, 1021, 1160, 1165, 1182, 1204, 1231, 1292, 1363, 1364, 1405

Risikobevölkerung, z. B. 505

Risiko, relatives: see relative Risk

Risiko-Theorie 64, 528, 864, 1204

Risk:
  see also risk measurement, estimation of risks
  analysis: systems reliability engineering 378, 524
  and insurance 64, 431, 528, 1300
  and risk factors: longitudinal models for chronic disease risk: an evaluation of logistic multiple regression and alternatives 1421
  assessment:
    and reliability engineering 378, 524
    of large technological systems 1165
    through the determination of rare event probabilities 1160
    weak association between an exposure and a disease 1429
  attributable 137, 338, 367, 1363, 1364, 1429
  classification in life insurance, e. g. 243
  estimating the population attributable … for multiple risk factors using case-control data 151
  estimation:
    see also cancer: avoidable risks
    assessment of acceptability, and the value of human life and safety 638

factors:
  and confounding, e. g. 682
  identification of, e. g. 137, 682, 770, 1182, 1314
  multiple, using case-control data, estimating the population attributable risk for 151

functions, general relative, for case-control studies 139

judgement about 361, 638, 1231

measurement, estimation of risks 84, 137, 198, 225, 243, 332, 361, 375, 418, 524, 528, 561, 569, 638, 682, 864, 903, 1021, 1160, 1165, 1182, 1204, 1231, 1292, 1363, 1364, 1405

of coronary heart disease, e. g. 137, 1314, 1360

of developing a disease: the description of the relation between risk and risk factors by logistic regression, e. g. 682, 1182, 1314

premium, e. g. 575

rare event probabilities 1160

rating the, e. g. 1231

ratio:
  and odds ratio 6, 86, 113, 137, 194, 367, 561, 569, 574, 682, 1182, 1204, 1292
  Bayesian inference on the odds and 569
  conditions for confounding of the 113
  crude and standardized, for k risk groups 386

Risks:
  competing, e. g. 234, 642
  disease, multifactor-multilevel, interaction contrast disease rates for assessing synergism or antagonism in 1091
  multifactor-multilevel disease risks 1091
  of cancer in the Untied States, quantitative estimates of avoidable, e. g. 292

Risks:
of contaminated air and drinking
water, of prenatal exposures,
and risks of energy sources 84
of drugs 620
Risk theory 64, 528, 864, 1204
elementary, e.g. 575
Rival models, discriminating among
165, 1046
Robot statistician 478
Robust:
and distributionfree procedures
214–217, 658
confidence intervals for location,
e.g. 547
covariance and correlation
matrices, e.g. 580
estimation:
and robust estimators 35, 253,
417, 488, 547, 580, 587, 828,
1209, 1247, 1248
procedures in multiple linear
regression, e.g. 273, 929
experimental design 1257
linear regression, e.g. 147, 273, 300,
547, 580, 587, 641, 929
line estimation 147, 273, 300, 547,
580, 587, 641, 929
location:
estimators 488, 547, 580, 1097
estimators in pairwise multiple
comparisons 314
Robustheit und robuste Verfahren,
robustness and robust procedures 123, 126, 214–217, 314,
393, 417, 520, 547, 580, 587, 663,
756, 830, 915, 916, 1040, 1123,
1257, 1323, 1358
Robust multiple comparisons:
of correlation coefficients 317
of means 102, 203, 309, 314, 393,
698, 1116, 1288, 1397
Robustness:
computational aspects associated
with, e.g. 520
in statistics 35, 123, 126, 214–217,
314, 393, 412, 488, 520, 547,
580, 587, 663, 756, 774, 830,
915, 916, 1040, 1123, 1257,
1323, 1358
modern trends in the theory of 123,
488, 520, 547, 580, 587, 756,
1323
of four methods for estimating the
mean 989
of statistical methods and nonparametric statistics, e.g. 488,
547, 580, 1097
of validity and robustness of
efficiency, e.g. 314
study of … by simulation: particu-
larly improvement by adjustment and combination 756
see also outliers
Robust:
nonlinear regression, e.g. 587, 661
principle components and multivariate normality 853
regression 147, 219, 300, 417, 547,
550, 551, 580, 587, 641, 765,
829, 1339
in industrial experiments 765
scale estimators, e.g. 488, 547, 580,
1097
like the MAD, e.g. 547
smoothing 417, 756
statistical analysis of interlaboratory studies 1123
Rootograms 1319, 1320, 1339, 1345,
1359
Roots, characteristic, e.g. 54
Rosendiagramme 839
Rotatable designs, e.g. 224
Rotated factor plots, e.g. 1284
Rotation: varimax, quartimax, equi-
max, orthogonal with gamma,
direct oblimin, direct quartimin,
orthoblique, e.g. 221, 1284
Rückfangmethoden 103, 966, 1000,
1064, 1065, 1208, 1210, 1391
Rumors and news: diffusion models
50, 51, 53, 966
Rundreisen, Logistik der 293, 867,
1022
Rundungsfehler 1074

Sacrifices in the Old Treatment, a game theoretic approach 128

Säuglingssterblichkeit 161, 1079, 1255

Safety characteristics of nuclear power stations 432

Saisonale Schwankungen in ökonomischen Zeitreihen, z. B. 1265

Saisonbereinigungsverfahren, z. B. 43, 202, 942, 1265
    siehe auch Zeitreihen-Analyse

Salesman, travelling, problem 293, 333, 867, 1022

Sample:
    design checklist, e.g. 373, 1243
    geometry and random sampling, e.g. 634
    influence curve 219
    size:
        and matching 21, 1182
        and power calculation using noncentral F approximations 462
        for non-parametric multivariate tolerance regions 5
    sizes:
        for attribute sampling plans 781, 984
        for comparing two proportions 367, 984, 1058, 1152, 1153, 1182
        for comparing two very small binomial probabilities, e.g. 146
        for estimation and for tests of significance 205, 367, 406, 461, 462, 1045, 1151–1153, 1182, 1198, 1214, 1238, 1348, 1395
    surveys 36, 373, 621, 644, 645, 658, 727, 729, 820, 1061, 1137, 1238, 1403, 1424, 1431, 1433
        incomplete data in 248, 714, 820, 1161
        missing data in 820, 1125, 1137, 1424
        nonresponse in 205, 594, 645, 804, 820, 928, 1153, 1275
        nonresponse and imputation in 248, 714, 820, 1161
        of finite populations, the development of 35

Samples:
    and census: combined uses and designs 676
    nonrandom, e. g. 1237
    trimmed samples and Winsorized samples 1358
    two, independent or matched, of response curves: non-parametric analysis 708

Sampling:
    applied 98, 645, 658, 721, 727, 729, 1061, 1137, 1169, 1201, 1261, 1271, 1274, 1395, 1403, 1431, 1433
    biological populations 223
    by attributes 483, 781, 984, 1181
    closed populations: see capture-recapture models
    cluster: see cluster sampling
    design for environmental studies 447
    designs for time series 720
    distributions, e.g. 658, 769
    double, e.g. 621, 1181
    errors, e.g. 1343, 1424
    for auditors 1395
    frames 621, 1424
    inspection:
        and quality control 781, 1389
        by attributes 483, 781, 984, 1181
    in surveys 248, 621, 729, 1137, 1238, 1424, 1433
    lattice, e. g. 621
    monetary-unit 1395
    multiphase, e.g. 621
    multistage, e.g. 621
    plans 483, 781, 983
    probability 621, 729, 1274
    quota 820, 1061, 1201, 1261, 1431, 1433
    representative 729
    sequential: see sequential sampling

Sampling:
strategies: robustness and
efficiency 595
strategy guided by auxiliary
information incorporated in a
model 595
stratified 21, 302, 621, 729, 1061,
1137, 1201, 1274, 1395, 1433
sub-sampling, e.g. 621
survey 176, 248, 621, 644, 714, 820,
928, 1061, 1274, 1342, 1343,
1371, 1424
systematic 593, 729, 1395, 1431,
1433
theory 98, 729, 1061, 1137, 1192,
1261, 1274, 1431, 1433
unit and sampling frame, e.g. 621
Sand, the fascination of, and statistics
35
SAS, Statistical Analysis System 83,
155, 173, 206, 430, 736, 770, 786,
897, 1017, 1196, 1264, 1316,
1328, 1340, 1437
Scale:
and location differences in two-
sample comparisons: graphical methods in nonparametric
statistics 363
estimates, robust 547, 580
estimators: see also robust
estimators
estimators like the MAD, e.g. 547
k-sample nonparametric tests for
1044
nonparametric tests for 214, 215,
315, 719
parameters, tests for 214, 215, 315,
782
robust and resistant estimators/
measures 547, 916
shift detection in a random sequence at an unknown time
point 1287
Scales of measurement 12, 117, 506,
965, 998, 1120, 1341, 1436
and measures of association, e.g.
792

Scaling:
see dual …, multidimensional …,
and nonmetric multidimen-
sional …
a data matrix in a low-dimensional
Euclidian space, e.g. 513
dual, e.g. 965
multidimensional, see there
optimal, and alternating least
squares, the combination of,
e.g. 1436
or/and shifting curves with one
bend, e.g. 15
Scatter plots 201, 219, 249, 363, 502,
755, 828, 916, 1122, 1320, 1359
influence-enhanced 1296
of variable loadings with factors as
axes e.g. 1284
smooting … and graphical
methods for regression, e.g.
755
Schätztheorie, z.B. 66, 580, 587, 658,
769, 774, 880, 1071, 1090
Scheidungsstatistik 161, 1255
Scheinkorrelation und das
Simpsonsche Paradoxon 884
Schichtenbildung, z.B. 779
Schichtgrenzen:
bei mehreren Schichtungs-
variablen 302
Optimierung und Stichprobenumfänge 302
Schichtung:
einer Grundgesamtheit: Zahl der
Schichten und Aufteilung des
Stichprobenumfanges auf die
einzelnen Schichten 302
und Aufteilung, Optimierung in
der Stichprobentheorie durch
1191
und Blockbildung: eine Einfüh-
rung 1151
Schiefe-Korrektur für einseitige
Vertrauensbereiche für
Binomial- und Poisson-
Parameter 486
Schiefe Verteilungen 184, 203, 630,
960, 1040

Schips-Stier-Verfahren 1265

Schließende Statistik, z. B. 19, 43, 48, 121, 286, 1151–1153, 1238, 1352

Schneeberger-Ansatz zur Bestimmung mehrdimensionaler Schichtungsvariablen bei gleichzeitig optimaler Stichprobenentnahme 302

Schul- und Hochschulstatistiken 161, 1255

Schwangerschaftsabbrüche: Daten 161, 591, 1255

Science: methods, nature and status, e. g. 182

Scientific:
  hypotheses: see scientific method
  method 123, 182, 220, 229, 291, 303, 423, 525, 754, 882, 1024, 1066, 1318, 1321–1323, 1361, 1404, 1411
  writing 328, 379, 1263

Score functions and density-quantile functions, e. g. 1025

Scores, cutting, the use of … in selection procedures 1007

Scoring:
  for ordered categories, e. g. 21
  of multiple-choice tests: application of the subset selection theory 406
  techniques, e. g. 160
  the categories of an ordinal response variable: see ridit analysis

Screening 79, 88, 160, 225, 367, 419, 875, 940, 983, 1007, 1008, 1145, 1274
  compounds for clinically active drugs, e. g. 88, 160
  for the early detection of cancer 225
  in chronic diseases 910
  medical 79, 225, 367, 419, 1145

Seasonal adjustment:
  graphical methods for 202
  methods, e. g. 4, 958

Seasonal:
  and calendar adjustment 718

autoregressive integrated moving average models, e. g. 4

Seasonality:
  and seasonal adjustment of time series, e. g. 4, 97
  stochastic and deterministic trends, e. g. 4, 122, 1019

Seasonal models, e. g. 4, 97, 122, 718, 1019

Secondary storage methods for statistical computational operations 97

Secretary problem 1, 381, 406, 407

Secret messages: see cryptography

Segmentation of a sequence and of spatial data 513

Sekretärinnen-Problem 1, 381, 406, 407

Sekundärdaten – Nutzung in Sozialmedizin und Epidemiologie 133

Selbstmorde und Totgeburten: Trends 58

Selecting:
  a population-based control group by random digit dialing 497
  a subset of variables that contains as much information as possible 858
  good populations 307, 405–407, 465, 466
  the best process 405

Selection:
  and ranking procedures 74, 307, 406, 407, 465, 466
  bias in epidemiology 682, 1153, 1230
  non-random samples, e. g. 1237
  of composites 1323
  of independent variables in regression analysis 249, 300
  of principal variables 858
  of subsets of regression variables, e. g. 300, 890
    bias reduction 890
  of the smallest or largest entity 406
  procedures:
    subset: see subset selection

Selection:
  the robustness of, e.g. 1097
  the use of cutting scores in 1007
Selektion und Genetik 1093
Self-consistency in survival data
    analysis 234
Semigraphic displays 1319
Semilogarithmic charts, e.g. 1188,
    1189
Semi-Markov process, e.g. 233
Sending secret messages: see
    cryptography
Sensitive:
  data, obtaining, e.g. 311, 329, 820,
    1260
  questions: see randomized re-
    sponse techniques
Sensitivitätsanalyse in der linearen
    Programmierung: siehe Opera-
    tions Research
Sensitivitätskurve: siehe sensitivity
    curve
Sensitivität und Spezifität 2, 78, 79,
    86, 213, 250, 367, 390, 620, 682,
    986, 1355
Sensitivity:
  (in statistical inference) 1055, 1257
  and specificity (diagnostics) 2, 78,
    79, 86, 213, 250, 367, 390, 620,
    682, 986, 1355
  curve 488, 547, 580, 1097
  experiments, e.g. 286
Sensorik, sensorische Prüfung von
    Lebensmitteln 618, 1107
Separable programming, e.g. 1308
Separating mixtures of distributions
    224, 343, 869
Sequenced decisions 398, 654, 1119
Sequential:
  allocation rules 55
  analysis 28, 55, 286, 694, 769, 1348,
    1374, 1388, 1393
  and recursive estimation, e.g. 1071
  clinical trials 28, 55, 1393
  estimation, nonparametric 719
  experimentation: designs for 270,
    1257

medical trials 28, 55, 1393
  methods 286, 658, 694, 769, 1348,
    1374, 1388, 1393
  nonparametric tests 719
  probability ratio test 55
  sampling 483, 781, 1065, 1181
Sequentially rejective multiple test
    procedures 59, 565, 1180, 1242
Sequential unconstrained minimiza-
    tion technique applied to opti-
    mal systems reliability 1308
Sequentiell verwerfende Testproze-
    duren 59, 565, 1180, 1242
Serial correlation 300, 516, 658, 672,
    673
  Durbin-Watson test for 672, 673,
    1350, 1437
  in residuals, e.g. 300
  in univariate repeated measures
    analyses, e.g. 516
Serial measurements:
  and covariance structure, e.g. 1368
  linear models for the analysis of
    longitudinal studies, e.g. 1368
Seriationsproblem 596, 840
Series, some useful, e.g. 1028
Service:
  and level of analysis in statistical
    consulting 477, 888, 1149,
    1257, 1441
  systems 193, 339, 347, 695
Sexualproportion der Lebendgebo-
    renen, Entwicklung der 626
Shaffer's extension of Dunnett's
    procedure 787
Shape and level comparisons of two
    response curves 703, 708
Shapiro-Wilk:
  test for normality, e.g. 129, 808,
    853, 1390
  and Shapiro-Francia tests, e.g.
    1052, 808
Shift function, e.g. 363
Short stories on statistics 568
Shelf life of drugs for the zero-order
    model 160
Shrinkage estimation, e.g. 774

Shrinking factor: James-Stein estimator, e.g. 327

Sicherheit von Zeugenaussagen, z.B. 76

Side-effects in clinical trials, e.g. 160

Signal estimation and detection 718

Signed-rank procedures, e.g. 1072

Significance levels:
  combining and comparing ... from nonindependent hypothesis tests 414, 1132, 1153, 1270, 1362
  combining and contrasting 1270

Significance tests, e.g. 231, 572, 643, 655, 658, 769, 892, 1298, 1413

Significance tests, integrating, e.g. 414, 1132, 1362

Signifikanztests, z.B. 231, 572, 643, 655, 658, 769, 892, 1298, 1413

Signifikanztest und Bayessche Hypothesenprüfung, z.B. 48, 1147, 1298

Sign test 285, 709, 766, 836, 1152, 1153
  multivariate, e.g. 708

Similarity aggregation:
  see also clustering and cluster analysis
  methods, e.g. 612

Similarity:
  analysis, e.g. 612
  and distance matrices, graphical displays of 342
  and preference analysis 612
  see cluster analysis

Simplex-centroid design and simplex-lattice designs, e.g. 224

Simpson's paradox, e.g. 414
  and illusory correlation 884

Simulation 129, 130, 179, 424, 536, 661, 695, 759, 760, 849, 900, 908, 1013, 1034, 1112, 1143, 1168, 1190, 1192, 1375
  general purpose simulation system, e.g. 760, 849, 1033, 1375
  output data, statistical analysis of 759

Simulationsrechner und Simulationssprachen, z.B. 424

Simulation:
  studies regarding robustness of statistical methods: see robustness
  technische und insbesondere auch regeltechnische Anwendungen, z.B. 424
  with GPSS, e.g. 424, 760, 849, 1033, 1375

Simultaneous:
  equation models, e.g. 22
  statistical inference, e.g. 895

Single-case experimental designs to study behavior change 413, 527, 571, 1047

Single-subject designs, time series analysis for 413, 527, 571, 1047

Singularity and multicollinearity: remedies, e.g. 1284

Singular value decomposition technique 417, 661, 831

Sinuosity ratio, e.g. 1333

SIR 1196

Size:
  of a closed population, estimating the 103, 966, 1000, 1064, 1065, 1208, 1210, 1391
  of a sample, see sample size

Skalierungen, Theorie des Messens 12, 117, 506, 998, 1120, 1341, 1436

Skalierung:
  "kategorischer" Daten 12, 965
  nichtmetrische Modelle, Theorie der verbundenen Messung 1341

Skewness 184, 290, 630, 739, 960, 1040, 1152, 1153, 1390
  and kurtosis, bounds on the sample 630
  and kurtosis, measures of multivariate, e.g. 840

Slippage tests, e.g. 49

Slope:
  ratio assays 160, 359, 531, 582

Slope:
    testing, nonparametric ... in linear
        regression 436
Small effects in social research:
    combining results 1132
Smallest space analysis 969
Small numbers, law of, belief in the
    640
Smoothing:
    and decomposition time-series
        methods, e.g. 826
    and forecasting, e.g. 4, 107
        see also time series analysis
    data in exploratory data analysis
        201, 871, 1320, 1339
    robust 201, 417, 756
Soccer: distribution of goals per
    game 1063
Social:
    change, the cohort as a concept in
        the study of 851
    experimentation in the United
        States, large-scale 35
    medicine, bias reduction in 205
    mobility tables: pseudo-Bayes esti-
        mates of cell probabilities,
        e.g. 103
    models, e.g. 1113
    research, meta-analytic methods
        in, e.g. 414, 588, 1132, 1270
    statistics 51, 650, 913
        for comparative assessments 35
Sociological research 101, 117, 303,
    414, 588, 603, 845, 851, 1132,
    1246, 1270, 1361
Sociology:
    causal analysis in 101
    theory building 101, 303, 603
Software:
    and computing networks,
        reliability of 97
    building more intelligence into sta-
        tistical computer programs
        478, 507, 1346
    BMDP: see there
    EDA 871, 1339
    epidemiologic analysis with a pro-
        grammable calculator 1140

-failure data, analysis of, e.g. 46
for a linear and a nonlinear least-
    squares curve fitting program,
    e.g. 249
for a structural equation program,
    e.g. 926
for basic business statistics, e.g.
    786
for cluster analysis, e.g. 717, 1129
for efficiently computing the
    reliability of arbitrary
    planar networks, e.g. 46
for loglinear models, e.g. 114, 285,
    409, 428, 471, 472, 863, 864
for microcomputer 97, 404, 520,
    679, 741, 918, 933, 1423
for multivariate statistics, e.g. 93,
    114, 285, 376, 404, 661, 900,
    1078, 1412
for nonparametric survival data
    analysis 144, 770
for reliability analysis, e.g. 524
for seasonal exponential
    smoothing, e.g. 4
for statistical graphics, e.g. 69, 168
for statistical methods in cancer
    research, e.g. 888
for survey data analysis 1137, 1213,
    1433
for survival data analysis, e.g. 770
for time series analysis, e.g. 181
for "x": see under "x" and (under)
    statistical software
for univariate and multivariate
    analysis of variance, a
    comparison 716
for Winters' additive and multipli-
    cative seasonal forecast
    procedures, e.g. 4
FORTRAN: see there
für die statistische Analyse zensier-
    ter Daten, z.B. 1177
reliability, e.g. 46, 97
    growth model, inflection
        s-shaped 999
    growth, nonhomogeneous error
        detection rate models for
        999

SAS: see there
SPSS: see there
Statistik-Programmsysteme und Hinweise 83, 93, 114, 168, 285, 376, 404, 639, 661, 741, 786, 891, 897, 918, 926, 1001, 1121, 1122, 1129, 1284, 1339, 1346, 1412
survey of statistical, e.g. 376
and special systems 83, 93, 114, 168, 285, 376, 404, 639, 661, 741, 786, 891, 897, 918, 926, 1001, 1121, 1122, 1129, 1284, 1339, 1346, 1412
writing programs to analyze large data matrices 767
Somers':
delta, Somers' d, e.g. 9, 11, 724, 766, 1110, 1334
d and mean ridit, e.g. 611
Sonderbares aus der Mathematik 705
Sorting, multidimensional, e.g. 97
Soziale Schichtung 707
Sozialmedizin, Bias-Reduzierung in der 205
Sozialprodukt, z.B. 1171
Sozialstatistik (siehe auch Bevölkerungsstatistik) 726, 727, 1255
Sozialwissenschaftliche Methoden 117
Space and time, data collection for details over 1424
Spaces, linear Euclidean, and unitary, e.g. 749
Spatial:
and temporal analysis in ecology 222
autocorrelation coefficient: random, negative or positive correlation and their statistical significance, e.g. 1333
interpolation and smoothing of data 47
point patterns: analysis of 47
process and spatial-temporal process, e.g. 233
relationships between variables: see correspondence analysis

statistics 389, 968, 1118, 1333
Spearman-Brown formula, e.g. 739
Spearman-Rangkorrelation 317, 439, 597, 836, 1153, 1426
Spearman's:
rank correlation 317, 439, 597, 836, 1153, 1426
r_s, nomogram for obtaining z-test statistic from ... and sample size 10 to 1000 439
Specificity, diagnostic 2, 78, 79, 86, 213, 250, 367, 390, 682, 986, 1355
Spectra: cumulants and cumulant spectra 718
Spectral analysis:
in analytical chemistry 377
of economic time series, e.g. 122, 186, 285, 718, 777, 1080, 1323
robust-resistant 718
Spectral:
decomposition: see matrix algebra, e.g. 634
estimation, autoregressive 718
Spectrum:
analysis, use of fast Fourier transforms in 185, 186, 255, 718, 859, 1184
estimation, computer programming of 718
Speech recognition:
continuous, statistical methods 717
using LPC distance measures 720
Spektralanalyse ökonomischer Zeitreihen, z.B. 122, 185, 186, 285, 718, 777, 1080, 1184, 1323
Spezifität und Sensitivität: Weiterführendes 2, 78, 79, 250, 986
Spherical:
data and statistics 57, 364, 839, 840, 1029, 1373
normal distribution 31, 1373
Sphericity:
among repeated measurements, e.g. 975
and/or circularity, e.g. 44, 975
Spieltheorie 52, 89, 128, 212, 536, 637, 758, 1009, 1011, 1099, 1092, 1302

Spline:
  estimates, e.g. 1071
  functions in data analysis 1416
  regression 300, 341
Splines in statistics 1235, 1379, 1416
Split:
  middle trend estimation technique
    527
  plot design 251, 674, 896, 901,
    1036, 1238, 1408, 1422
  plots and/versus repeated
    measures designs 896, 901
Sprache: Hinweise zur deutschen
    Sprache 110
Sprachverschleierung: siehe Kryp-
    tographie
Spreadsheet software 171
Spread-versus-level plot, e.g. 547
SPSS and SPSS-X: programs and
    guides 83, 93, 404, 639, 736, 786,
    971, 972, 1195, 1196, 1249, 1250,
    1264, 1284, 1437
Square contingency tables:
  analysis of 11, 103, 470, 793, 836,
    837, 861, 1056, 1152, 1153,
    1334, 1354, 1449
  simultaneous sign tests for margi-
    nal homogeneity of 773
  tests for marginal homogeneity in
    11, 103, 367, 710, 773, 836,
    1056, 1449
  with ordered categories, e.g. 11,
    103, 861, 1449
Squared:
  multiple correlation, adjusted, e.g.
    1284
  ranks test 215, 315
  semipartial correlation 1284
Square mean root rule 406
S system 69, 70
Stability of drugs: accelerated storage
    tests 160
Stable portfolio analysis 1078
Stages, method of, e.g. 99
STAN, statistical analysis: micro-
    computer program for regres-
    sion models, e.g. 15

Standard deviation, e.g. 658
  its standard error, some other stan-
    dard errors in terms of s, in
    terms of the measure and the
    relative number of cases re-
    quired 739
Standard errors of percentiles and of
    various measures of variability,
    skewness, kurtosis, and of
    various correlation coefficients
    and differences 658, 739
Standardisierte Chiquadrat-Werte,
    z. B. 1384
Standardisierung:
  als Verallgemeinerung der Index-
    rechnung 340, 349, 1428
  von Maßzahlen aus Epidemiologie
    und Bevölkerungsstatistik 21,
    86, 103, 194, 340, 349, 357,
    367, 602, 677, 678, 791, 805,
    916, 1321, 1428
Standardization:
  siehe auch Standardisierung
  and analysis of covariance as in-
    stances of prediction: using a
    table of proportions on the
    logit scale 751
  direct and indirect, for comparison
    103, 340, 805, 916, 1321, 1428
  direct, and the general linear
    model: a comparison 805
  of a table of means on the raw or
    log scale, e.g. 805
  of a table of proportions on the
    logit scale, e.g. 805
Standardized:
  and adjusted residuals in two-way
    tables, e.g. 472, 1110
  change score analysis 585
  mortality ratio and related
    adjusted rates with their
    standard errors 194
  normal quantiles, e.g. 643
  residuals 1316, 1340
  tables, e.g. 1110, 1153
Standardizing for comparative pur-
    poses, do rates need ...? 357

Standardnormalverteilung:
  Approximationen 39, 1029
  Schranken 65, 1392
Standard setting for atmospheric car-
  bon dioxide, dynamic, a game
  theoretic approach 128
Standortauswahl und andere Bewer-
  tungen 324
Standorte: Logistik 294
STAR display, STAR glyph, stars
  351, 634, 1359
STATCAT 252
States:
  equilibrium states and non-
    equilibrium states in
    reliability, e.g. 420
  mean duration of individual and
    cumulated states, e.g. 99
State space diagrams, e.g. 99
Stationarity and diffusion, e.g. 454
Stationary and nonstationary models
  for forecasting and control, e.g.
  122, 718, 720, 1019
Statistical analysis:
  introduction to 121, 286, 1153,
    1238
  of repairable systems failure data
    32
  systems and database/methodbase
    systems 478, 507
  systems: data structures, stepwise
    reduction of 507
Statistical:
  and probabilistic thinking in
    courtroom settings 299, 354,
    356, 959, 1241
  assessments as evidence 299, 354,
    356, 959, 1241
  computing 7, 15, 24, 35, 83, 93, 106,
    114, 197, 206, 249, 252, 285,
    376, 404, 478, 508, 520, 639,
    661, 679, 736, 741, 786, 824,
    891, 897, 918, 925, 926, 945,
    972, 1001, 1085, 1121, 1122,
    1129, 1137, 1213, 1249, 1250,
    1284, 1339, 1346, 1412, 1423,
    1433
    see also matrices/matrix and
      software
    applied statistical algorithms,
      e.g. 453
    new developments in 404, 478
  conclusion validity and other types
    of validity 220, 1404
  consultant, general automated 478
  consulting 350, 477, 478, 888, 1149,
    1257, 1441
  context aided testing system,
    STATCAT 252
  decision theory 48, 87, 757, 758,
    1120
  density estimation 47, 401, 658,
    1288, 1386
  distributions, e.g. 454, 504, 541,
    542, 631, 658, 701, 768, 769,
    1028, 1030, 1320
    dictionary and classified biblio-
      graphy of 1030
    in ecological work, e.g. 995
  expert systems, e.g. 478
  graphics: see graphical displays,
    graphical methods and plot-
    ting
  inference 121, 229, 231, 304, 817,
    1153, 1209, 1238, 1322
    and the binomial distribution
      1072
  maps. e.g. 351, 1189
    and framed-rectangular chart,
      e.g. 200
  matching:
    and/versus imputation for
      nonresponse 248, 1125
    missing data and imputation
      248, 1125
    theory, alternatives and simula-
      tion 1125
  methods:
    see also robustness
    bibliographical hints 701, 1153
    for agriculturists 1036, 1159
    for comparative studies, e.g. 21
    for digital computers 1085
      see also statistical computing
        and Programmsysteme

Statistical:
  methods:
    introduction to 121, 286, 1153, 1238
    modeling, e.g. 121, 288, 830, 1215
    using quantile functions, e.g. 547, 1024, 1025
  models: see models
  packages:
    see/siehe Programmsysteme
    the evaluation of software packages 1407
  practice in bureaucracies 846
  problem solving, e.g. 192, 423, 1194
    see also statistical consulting
  programs: see/siehe Programm-systeme
  sciences: trends 478, 499, 532, 1321–1323
  significance tests: see significance tests
  software 83, 93, 114, 168, 285, 376, 404, 639, 661, 741, 786, 891, 897, 918, 926, 1001, 1121, 1122, 1129, 1265, 1339, 1346, 1412
    see also Programmsysteme and Microcomputer
    applied statistics algorithms, e.g. 453
    a survey of 376
    benchmark data sets for the flexible evaluation of 1407
    more intelligent … and expert systems, e.g. 478
    progress and prospects, e.g. 97
  sources and techniques 1106
  tables 365, 406, 496, 712, 719, 793, 814, 921, 1038, 1152, 1153, 1209, 1426
  terms, e.g. 701
    dictionary of 657, 991
  tests, tests of hypotheses 231, 572, 643, 655, 658, 769, 892, 1298, 1413
  theory 643, 658, 769, 830, 1055, 1072, 1090
Statistician as an expert witness 959

"Statisticians and liars" 568
Statistics:
  algorithms, applied, e.g. 453
  and model building, see there
  and nuclear energy, e.g. 432
  and philosophy, e.g. 675
  and public policy, e.g. 913
  and tennis, e.g. 407, 1062
  and the law 299, 354, 356
  bibliographical hints 701, 1153
  bureaus 846
  contrasted with other sciences, e.g. 675
  definitions 701, 1153
  descriptive: see e.g. 701
  dissertations in 1331
  encyclopedia of statistical sciences 701
  for engineers: a guide to books 480
  for environmental biologists 447
  general principles, e.g. 229, 236
  in agriculture 1036, 1159
  inductive, e.g. 121, 231, 291, 304, 473, 1153, 1238
  in ecology 995, 1065, 1208
  in engineering: see the special key-word or subject heading
  inferential, e.g. 48, 121, 286, 1153, 1238
  in geography 389, 968, 1118, 1333
  in medicine: see the special key-word or subject heading
  in ornithology 909
  nonparametric: see there
  of directional data 57, 364, 839, 840, 1029, 1373
  on spheres 57, 364, 839, 840, 1029, 1373
  order 214, 253, 495, 658, 793, 836, 1209
  resistant, resistance in statistics 502, 547, 1323
  robust 123, 126, 214–217, 314, 393, 520, 547, 580, 587, 663, 756, 830, 915, 916, 989, 1040, 1123, 1257, 1323, 1358
  social, recent developments in, e.g. 51

spatial 389, 968, 1118, 1333
summary: converting into correlation coefficients, e. g. 414
theory of 587, 643, 658, 769, 830
Statistik:
    beschreibende, z. B. 43, 81, 349, 1151–1153, 1352, 1374
    Daten zur Bundesrepublik, siehe dort
    deskriptive, z. B. 43. 81, 349, 1374
    des produzierenden Gewerbes 20, 268, 1255
    enzyklopädische Darstellungen siehe Enzyklopädie der Statistik
    Formelsammlung und Definitionen, z. B. 1352
    für Geographen 389, 968, 1118, 1333
    grundsätzliche Bemerkungen, z. B. 229, 236
    Methoden und Ziele 121, 658, 701, 1096, 1151–1153, 1238, 1244, 1318–1323, 1351
    -Programme für Kleincomputer 404, 741, 918, 933, 1423
    -Programm-Systeme 83, 93, 114, 285, 376, 404, 478, 639, 661, 741, 786, 891, 897, 918, 926, 1001, 1121, 1122, 1129, 1213, 1284, 1346, 1412
    schließende, z. B. 19, 43, 48, 1151–1153, 1238, 1352
    -Software: siehe Software: Statistik-Programm-Systeme
    umfassende Darstellungen, z. B. 658, 1096
    und Rechtsprechung, z. B. 959
Statistische:
    Analysesysteme 507, 520
    Beratung 477, 478, 888, 1149, 1257, 1441
    Datenbanken, Outputkontrollen zur Sicherung 1185
    Inferenz für lineare Parameter 587
    Klasseneinteilung 1070
    Methoden 121, 658, 701, 1096, 1151–1153, 1238, 1244, 1318–1323, 1351
    einführende Übersichten 121, 1151–1153, 1238, 1244, 1318–1323, 1351
    Modelle, z. B. 538, 1046, 1096
    Tests 231, 572, 643, 655, 658, 769, 892, 1298, 1413
Statistisches Bundesamt: Aufgaben 487, 554, 977
Statistische:
    Tabellen 365, 406, 496, 712, 793, 921, 1038, 1152, 1153, 1426
    Verteilungen z. B. 504, 541, 542, 631, 658, 701, 768, 769, 1028, 1320
Steady-state simulations, confidence intervals for: fixed-sample-size procedures, sequential procedures, and a replication-deletion approach 760
Stein estimation, e. g. 22, 175, 327
Steins's paradox, e. g. 327
Stem-and-leaf display or plot 502, 547, 871, 1319, 1320, 1339, 1358, 1359
Stepwise regression 300, 549, 550, 1383
    versus hierarchical clustering 442
Sterbestatistik 161, 166, 340, 1171, 1255
Sterbetafel:
    siehe Sterbetafel-Studien
    ansatzweise Berechnung der Kohortensterbetafel für Deutschland 282
    einfache Beispiele, z. B. 268
    -Studien 161, 194, 228, 282, 332, 340, 505, 667, 761, 1065, 1226, 1232, 1378, 1428
    und Entropie 619
Sterblichkeit:
    in Perioden- und Kohortenbetrachtung 194, 282, 1428
    nach Todesursachen 1148, 591, 1079, 1255
Sterology, recent trends in 619

Stichproben: Anteilschätzungen in
    Musterstichproben, z. B. 1169
Stichprobenauswahl 36, 98, 727, 729,
    1061, 1169, 1201, 1261, 1274,
    1431, 1433
    siehe auch Erhebungen
Stichproben bei Inventuren 302,
    1172, 1179
Stichprobenergebnisse, ihre
    Genauigkeit 727
Stichproben:
    Erhebung 36, 373, 727, 729, 1061,
        1403, 1431, 1433
    für Befragungen von Haushalts-
        mitgliedern, z. B. 1169
    für Bevölkerungsumfragen, z. B.
        1169
    geschichtete 21, 98, 302, 621, 729,
        779, 1061, 1137, 1169, 1201,
        1274, 1433
    Stichprobeninventur 302, 1172,
        1179
    Klumpenstichproben 98, 205, 621,
        729, 779, 1061, 1137, 1274,
        1431, 1433
    mehrphasige, bzw. mehrstufige,
        z. B. 779
Stichprobenpläne:
    Muster-Stichproben-Pläne, z. B.
        1169
    zur Qualitätskontrolle 460, 1329,
        1389
Stichprobentheorie 98, 729, 779,
    1061, 1137, 1201, 1261, 1274,
    1431, 1433
Stichprobenumfänge:
    für den Vergleich zweier sehr klei-
        ner Binomialwahrscheinlich-
        keiten 146
    für Schätzungen und statistische
        Tests 205, 367, 406, 461, 462,
        779, 1045, 1151–1153, 1198,
        1214, 1238, 1348
    zur Schätzung von Mittelwert-
        Differenzen 461, 462, 712,
        1045
Stichproben-Verfahren 98, 117, 658,

727, 729, 779, 1061, 1169, 1201,
    1261, 1274, 1431, 1433
Stigmatizing questions: see random-
    ized response techniques
Stirling numbers, e. g. 1028
Stochastic:
    approximation 719
    modeling of communication
        processes 900
    models:
        and processes 50, 53, 122, 193,
            233, 347, 454, 695, 874,
            1092, 1124, 1136
        and their forecasting, e. g. 122
        for mobility in closed social sys-
            tems, e. g. 50
        in reliability theory, e. g. 999
        of growth, e. g. 178, 1065, 1208
    ordering 1136
    order relations, e. g. 1136
    point processes of failures 32, 1304
    processes 53, 193, 347, 454, 874,
        1136
        Bayesian inference for, e. g. 53
        demographic, e. g. 193
        nonparametric inference for,
            e. g. 53
        sequential inference for, e. g. 53
    time series models/processes, e. g.
        4, 122, 185, 186, 718, 720, 958,
        1080
    transitional probability matrix, e. g.
        99
Stochastische Modelle und Prozesse
    50, 53, 193, 233, 347, 454, 695,
    874, 1092, 1124, 1136
Stock price movements, e. g. 50
Stop-loss premiums, bounds on 431
Stopping-rule bias in best subset
    selection 890
Storage:
    methods: see secondary ...
    tanks, e. g. 524
Stouffer's formula for combining
    significance levels, e. g. 1270
Straight-line:
    calibration problem: see
        calibration

fitting with both variables in error 147, 300, 547, 580, 587, 641, 832

Straightness plot: see transformations to linearize data

Strategische Spiele: siehe Spieltheorie

Stratification/Stratifizierung, e. g./z. B. 21, 621, 645, 682

Stratification:
see also survey and opinion research
criteria in classification of reference populations 441
for audit purposes 1395
versus analysis of covariance 674

Stratified:
analysis in epidemiology, e. g. 682
and frequency matching 21
incidence or mortality data analysis 137, 367, 561, 682, 1182, 1292, 1447
sampling 21, 302, 621, 729, 1061, 1137, 1201, 1271, 1274, 1395, 1433

Stress-strength models 265

Streudiagramme 249, 502, 828, 916, 1122, 1320

Streuungsunterschiede: Tests auf fehlende 215, 315, 1044

Strike duration modelling 763

Structural analysis:
see latent structure analysis, principal components analysis, and factor analysis
of discrete data 833

Structural equation:
modeling 169, 220, 344, 358, 628, 848, 926, 1404
BMDP program 926

Structural:
escalation, different aspects and multidimensional methods of, e. g. 829
modeling, e. g. 662
models in longitudinal-developmental investigations, statistical estimation of 954

Structure:
-activity relationships in pharmaceutical studies, e. g. 88
latent structure analysis 17, 425, 603, 814, 1078, 1347

Stuart-Maxwell-Test, e. g. 367

Stuart test of equality of marginals in a square contingency table, e. g. 836, 1449

Studentized maximum of equally correlated normal random variables, on the distribution of the 467

Student's t distribution: see t distribution

Studentized:
maximum modulus distribution 68, 1269
range distribution and Bonferroni's inequality 31
residuals 249, 300, 548, 1264, 1383

Student-Newman-Keuls test: see Newman-Keuls-Student test

Studies:
classifying, e. g. 414
integrating:
errors and limitations in, e. g. 414
research findings across, e. g. 414, 588, 1132, 1270

Subgroups:
building homogeneous subgroups on the basis of the similarities of objects: see cluster analysis
fallacies in the analysis of, e. g. 588

Subjective probability:
see also Bayes-Verfahren
and fuzzy events or fuzzy sets 464, 1367, 1439, 1445
theories 498

Subjektive Verhältnisse: Versuch ihrer Objektivierung 76

Sub-sampling, e. g. 621

Subset selection:
of regression variables, e. g. 300, 549, 550, 890
procedures 307, 406, 466, 467

Subsets of regression variables, selection of, e.g. 300, 549, 550, 890
Sufficiency, e.g. 66, 658, 1072
Superpopulation models for sample surveys with nonresponse 804, 805
Superposition of independent point processes, e.g. 233
Suppressor variables, e.g. 210, 1284
Sure-thing principle in decision theory and the reversal paradox 884
Surfaces on maps, e.g. 1333
SURVAN, a nonparametric survival analysis package 144
Survey:
  analysis: computer programs 1137, 1213, 1433
  and opinion research 36, 176, 196, 205, 248, 278, 329, 367, 422, 558, 564, 594, 621, 644, 676, 727, 805, 820, 845, 872, 992, 1050, 1137, 1161, 1200, 1243, 1262, 1274, 1275, 1276, 1342, 1371, 1382, 1403, 1424, 1433
  assessment of errors in household surveys in developing countries 1342, 1343
  by telephone versus personal interviews 384, 459, 497, 1200, 1276, 1325
  data:
    analysis, software for 1213, 1433
    application of the general linear model to, e.g. 1137
    collection, planning and management, e.g. 1137
    design checklist, e.g. 373, 1243
    errors and other limitations, e.g. 1424
  organizations, management of, e.g. 1137
  questions and how they influence the answers 646, 1200, 1424
  research:
    causal modeling, e.g. 1137
    design and data analysis, e.g. 1137

  using a telephone 384, 459, 497, 820, 1200, 1260, 1276, 1325
  sampling 176, 248, 595, 621, 644, 645, 714, 820, 928, 1061, 1274, 1342, 1343, 1371, 1424
    approaches to enumerative, e.g. 820
    robustness: the asymptotic and the prediction approach 595
Surveys:
  as social indicators: problems in monitoring trends, e.g. 1137
  criticism of, e.g. 845
  data collection for details over space and time, e.g. 1424
  error profiles: uses and abuses, e.g. 1424
  hot-deck procedures 820
  household, assessment of errors in 1342
  imputation in 248, 714, 820, 1125, 1161
  incomplete data 248, 714, 820, 1161
  integration of 872
  measurement in, e.g. 1137
  missing data in 804, 820, 1125, 1137, 1424
  nonresponse in 205, 594, 804, 820, 928, 1153, 1275
  nonsampling errors in 594, 714, 727, 1424
  outlier detection and data editing, e.g. 1424
  pattern recognition techniques, e.g. 1424
  response effects, e.g. 1137
  response error reduction 167, 594, 992, 1275, 1424
  sample: see sample surveys
  using exploratory data analysis:
    as an aid in modeling and statistical forecasting, e.g. 1424
    to monitor socio-economic data quality, e.g. 1424
Survival analysis:
  Cox models for 228, 234, 285

in operations research, e.g. 974

Survival data analysis: 21, 134, 155, 173, 184, 193, 225, 234, 242, 285, 332, 437, 625, 649, 694, 761, 770, 864, 888, 893, 950, 952, 974, 994, 1060, 1198, 1226, 1291

see also proportional hazard estimation

bivariate survivor functions 234

censored survival data: see censored …

goodness of fit, e.g. 893

nonparametric, a computer program 144, 770

one sample nonparametric methods, e.g. 234, 893

probability plots and hazard plots 952

regression analysis of survival data in cancer chemotherapy 173

survey of related topics 952

two and k sample nonparametric methods, e.g. 770, 893

with nonparametric techniques, e.g. 234, 770, 893

with the logrank test, e.g. 21, 694, 893, 1291

Survival data:

dependence on explanatory variables 234

graphical methods 952

models for, e.g. 193, 234, 332, 770, 864, 952

time-dependent covariates 234

Survival:

dependence on explanatory variables 234

distribution fitting, the Gehan-Siddiqui regression method for, e.g. 770

distributions:

see also survival data analysis

parametric and nonparametric methods for comparing two survival distributions 770

testing the equality of, e.g. 234, 625, 694, 770, 893, 952, 1291

function of cohorts and period life tables 194, 282, 1428

Kaplan-Meier estimation of: see Kaplan-Meier's estimator

in cancer therapy 173, 437, 649, 888, 1060

prediction:

from regression models, e.g. 994

see also survival data analysis and other survival entries

rates:

of patients: comparing the survival of two or more groups of patients adjusted for normal mortality expectation 145

relative: comparing the survival of two or more groups of patients adjusted for normal mortality expectation 145

relative five-year, and corrected, e.g. 770

studies:

cancer: examples 437, 649, 1060

correspondence analysis in 612

logrank test, e.g. 21, 694, 893, 1291

nonparametric frequentist proposals for monitoring comparative 719

times, analysis of 134, 155, 173, 184, 234, 242, 332, 437, 625, 649, 694, 761, 770, 864, 888, 893, 950, 952, 974, 994, 1060, 1198, 1226, 1291

trials and other medical trials, correspondence analysis in 612

Survivor functions, e.g. 99, 770

bivariate, e.g. 234

of systems, e.g. 420

Suspended rootograms: see there

SWEEP operator, e.g. 15

Switch in a regression model, tests proposed to deal with the problem of a, e.g. 1087

Switching point or points in a regression model, e.g. 1087
Symbolic:
computing and statistics, e.g. 520
differentiation and optimization of multivariable functions, e.g. 445
Symbolik: mathematische Zeichen 279
Symmetrical distributions, long tailed 1127
Symmetric random walks, e.g. 643, 1136
Symmetrie:
Graphische Methoden und Tests 290
mehrerer abhängiger Stichproben 94
Tests, die auf der Prüfgröße von Watson basieren 534
-Tests für Dreiweg-Kontingenztafeln 1101
zweier abhängiger Stichproben 709
Symmetry:
and marginal homogeneity in square contingency tables 103
compound, as one form of sphericity, e.g. 975
hypothesis of 719
in hierarchical data, a distribution-free test for 94
plots and tests 290
quantile plots for assessing 363
tests:
of … based on Watson statistics 534
of … in three-way contingency tables 1101
System analysis:
in sociology e.g. 303
and time series 1019
Systematic:
errors 21, 172, 205, 367, 506, 568, 640, 682, 726, 1074, 1182
measurement errors 733, 1081, 1199, 1387, 1435

sampling 593, 645, 729, 1395, 1431, 1433
Systematische Fehler:
Bias: ihre Vermeidung 21, 172, 205, 367, 506, 568, 640, 682, 726, 1074, 1182
in der Psychologie und ihre Vermeidung 506
in wirtschafts- und sozial-statistischen Daten, der 726
Systeme:
nichtreparierbare und reparierbare, z.B. 32, 99, 265, 378, 420, 552, 1168, 1304, 1327
und Modelle, z.B. 1411
System, knowledge base management, e.g. 1108
Systemorganisation, Theorie der 410, 529, 1033, 1104, 1105
Systemplanung, z.B. 1033
System reliability evaluation, e.g. 32, 99, 420, 847
Systems:
analysis and modeling in defence 581
comparing alternative, statistical techniques for 760
data base 104, 576, 1330, 1349
data- and methodbase, e.g. 507
dynamics, e.g. 522
having a decreased hazard rate: statistical estimation for reliability improvement, e.g. 265
Systemsimulation, z.B. 1033
Systems:
maintained, availability of, e.g. 847
modeling and simulation: see simulation
Monte Carlo investigation of, e.g. 420
new, time to first failure, e.g. 420
of linear equations, e.g. 54, 443, 1085, 1207
organization using Petri nets 410, 529, 1104, 1105

reliability:
   and risk analysis, e.g. 378
   interval estimates, multistate
      systems, repairable
      systems, e.g. 265
   optimal:
      see discrete maximum princi-
        ple, generalized reduced
        gradient method, general-
        ized Lagrangian functions
        method, Lagrange multi-
        pliers and Kuhn-Tucker
        conditions, and sequential
        unconstrained minimiza-
        tion technique
      heuristic methods applied to
        1308
   optimization of 1308, 1327
   using Bernoulli sampling of
      components, e.g. 265
   selection of 760
   series, parallel, series-parallel, par-
      tially redundant and standby
      redundant systems, e.g. 99
   survivor functions, e.g. 234, 420,
      770
Systemtechnik 1033, 1165, 1308
Systemverbesserung, z.B. 1033

Tabellen der mathematischen Statistik
   365, 496, 684, 712, 793, 814, 921,
   1038, 1152, 1153, 1426
Tabellen-Gestaltung, z.B. 1151
Table of percentage points of the
      range of independent t variables
      1397
Tables:
   and their normalization 576
   for testing multivariate statistical
      hypotheses 712, 814, 1038
   helpful in acceptance sampling,
      e.g. 1181
   helpful in statistical inference 365,
      496, 684, 712, 719, 793, 921,
      1038, 1152, 1153, 1426

Tariff games 128
Tau (measures of association) 81, 349,
   429, 724, 766, 836, 1110
Taylor-Russell tables 739
t charts 914
t distribution:
   algorithm to obtain critical values
      191, 661
   approximations 158, 400, 693
   exact p values for t tests 744
   geometry of the t statistic and the
      T^2 statistic 634
   inverse, a retrievable recipe for 400
   multivariate: on the distribution of
      the Studentized maximum of
      equally correlated normal
      random variables 467
   multivariate, with zero correla-
      tions, percentage points of
      188
   noncentral 541, 633, 1005, 1028
   percentage points 158, 191, 661,
      693
   probabilities 661
   table for simultaneous control on g
      contrasts 391
T^2 distribution, Hotelling's, e.g. 22,
   634, 800, 840, 1089
Teaching:
   -consulting interaction, e.g. 1149
   experimental design, e.g. 1257
   statistics, e.g. 1149
Technology and pure science, on the
      interplay of, e.g. 1113
Telefonzellen-Auslastung 280
Telephone:
   households, sampling of 384, 459,
      497, 820, 1200, 1276, 1325
   interview 384, 459, 492, 497, 820,
      1200, 1260, 1276, 1325
      application of the randomized
        response model, e.g. 1260
   survey versus personal interviews
      384, 459, 497, 1200, 1276, 1325
Temporal and spatial analysis in
      ecology 222
Tennis statistics 407, 1062

Terms used in:
  epidemiology, e. g. 682, 753
  statistics, e. g. 657, 701, 991
Terpstra-Jonckheere-Test: siehe
  Jonckheere-Test
Test:
  factor standardization, e. g. 805
  for non-additivity, e. g. 240, 674,
    732
Testing:
  homogeneity of dependent
    samples 709
  hypothesis, e. g. 48, 59, 66, 229, 231,
    572, 643, 655, 658, 769, 892,
    1298, 1413
  nondestructive, statistical analyses
    for, e. g. 265
  non-normality 129, 290, 781, 806,
    821, 1152, 1153
    in quality control 781
  the equality of two independent
    chi-square values 263
Testitems und Testskalen, z. B. 117
TESTRAND 308
Tests:
  bei Normalverteilung und bekann-
    tem Variationskoeffizienten
    267
  combination of independent 719
  of dominance for sequential
    categorical data 1365
  of hypotheses and test theory 231,
    572, 643, 655, 658, 769, 892,
    1298, 1413
  of significance 231, 572, 643, 655,
    658, 769, 892, 1298, 1413
  permutation 322, 629, 1072
Test theory and tests of hypotheses
    231, 572, 587, 643, 655, 658, 769,
    892, 1298, 1413
Tetrachoric correlation, e. g. 210, 276,
    284, 427, 739, 798
Theils Ungleichheitsmaß, z. B. 349
Theory:
  building 182, 303, 423
  of error 494, 726, 1102, 1342, 1343
  of statistics 66, 587, 643, 658, 769,
    774, 830, 1055, 1072, 1090

Therapiestudien 96, 694, 888, 1348
  Planung von Erhebungsbogen und
    Datenbank 776
  Zwischenauswertungen von 692,
    694
Therapeutic studies 96, 694, 888, 1348
Third variable analysis, e. g. 1277
Thomas distribution, e. g. 542
Three-group resistant line methods,
    e. g. 547, 1339
Three-parameter lognormal distribu-
    tion, e. g. 207, 208, 452, 647, 735
Tierversuch: Normalbereiche, z. B.
    1353
Tilted graph 1320
Time:
  -dependent covariates in survival
    data analysis 234
  domain models, e. g. 720, 1080
  models for closed and open social
    systems, e. g. 50
  for a disease to lead to recovery or
    death and duration of strikes
    763
  order detection 635
  series:
    see also ARIMA- and ARMA-
      models and processes
    see also economic time series
    analysis 4, 20, 24, 43, 93, 114,
      122, 140, 169, 177, 185, 186,
      202, 220, 230, 255, 274, 312,
      413, 444, 500, 518, 523, 571,
      626, 650, 658, 672, 673, 718,
      720, 769, 777, 778, 859, 865,
      906, 928, 942, 953, 958,
      1019, 1041, 1080, 1085,
      1121, 1171, 1184, 1203, 1265
    see also Zeitreihen-Analyse
    adaptive filtering, e. g. 811
    application of the Levinson
      algorithm 906
    at an advanced level 122, 140,
      185, 186, 230, 255, 285, 518,
      650, 658, 718, 720, 778, 859,
      865, 906, 958, 1121, 1184,
      1265

Bayesian models, e.g. 143
forecasting: see there
for single-subject designs 413, 527, 571, 1047
interrupted 500, 865
model selection techniques in 720
of error-correction models, e.g. 650
recent developments 230, 518, 958
Toeplitz matrices in, e.g. 54
with unequally spaced data 720
and pooled cross-sectional data 274
and system analysis, e.g. 1019
ARIMA and ARMA models: see there
autoregressions, robust estimation for 756
data:
    and pooled cross-sectional data 274
    frequency-domain analysis of multidimensional 718
    one-way analysis of variance with 1432
did an event have an impact on a …? 865
discriminant analysis for 717
economic, the econometric analysis of 523, 650, 777, 953, 1265
estimation of parameters in dynamical systems 720
experiments, design and analysis of, e.g. 413
impact patterns of external interventions on 865
in psychology 413, 444, 574
in social sciences, e.g. 859
in surveys, e.g. 422, 928
interrupted 500, 865
in the frequency domain 718, 1080
in the time domain 720, 1080

likelihood ratio tests on covariance matrices and mean vectors of complex multivariate normal populations and their applications in 718
measuring attenuation 720
models: non-linear … and dynamical systems, e.g. 720
nonparametric testing for 312
nonstationary, e.g. 720
    autoregressive, e.g. 720
of age-specific marital fertility schedule, analysis of age, period and cohort effects in 851
Package 181
plots, e.g. 871
regression, replicated: signal estimation and detection 718
sampling designs for 720
seasonality and seasonal adjustment of, e.g. 4, 97, 122, 718, 1019
segmentation of 650
spectral analysis 122, 186, 285, 718, 777, 1080, 1323
nonstationary and stationary models for forecasting and control, e.g. 122, 1019
time domain models, e.g. 720, 1080
space clustering of disease 225, 682
-structured data, uni- and multivariate analysis of variance of 954
-to-failure distribution:
    based on the normal distribution, e.g. 420
    determining the, e.g. 420, 974
-to-failure models: see survival data analysis, lognormal distribution and Weibull distribution
t, inverse, a retrievable recipe for 400
TNM-Klassifizierung und -Validierung 692

Todesrate 194, 340, 966, 1255, 1428
Todesursachen-Statistik 166, 591,
    818, 1148, 1255
Toeplitz matrices in time series
    analysis, e.g. 54
Tolerance:
    intervals for binomial and Poisson
        variables 479
    limits for sample variances 1306
    regions 5, 479, 485, 622–624, 1072
        bivariate normal 485
        for relative errors 624
        multivariate, designing the sam-
            ple size for nonparametric
            5
        sample size for the non-para-
            metric multivariate case 5
Toleranzfaktoren:
    für die Normalverteilung 921, 983,
        1426
    für die zweidimensionale Normal-
        verteilung 485
Toleranzgrenzen, verteilungsfreie,
    z.B. 1426
Toleranzintervalle für normalverteilte
    Meßfehler 624
Toleranzregionen 5, 479, 485,
    622–624, 1072
Topologically distinct channel
    networks, e.g. 1333
Totgeburten und Selbstmorde:
    Trends 58
Touren und Rundreisen, Logistik der
    293, 867, 1022, 1052
Tournament statistics 406
Toxikologische Standardversuche
    1353
Transfer function:
    analysis, e.g. 4
    model building, e.g. 122
Transferfunktion-Realisierung 1265
Transformation: expected normal
    scores 943
Transformationen 216, 217, 547, 716,
    916, 943, 955, 956, 1152, 1153,
    1320, 1390, 1426
    für Prozentzahlen und für kleine
        Anzahlen 799

Transformation:
    of various summary statistics into r
        414
    plot:
        for additivity, e.g. 547
        for equal spread, e.g. 547
        for symmetry, e.g. 547
        for the two-way table, e.g. 547,
            1320
    ridit 367, 611, 692
Transformations 216, 217, 547, 716,
    916, 943, 955, 956, 1152, 1153,
    1320, 1390, 1426
    in regression models 4, 34, 83, 300,
        547, 1001, 1098, 1153, 1383
    matched, e.g. 547
    power, e.g. 34, 547, 716, 916, 1320,
        1390
    scaling or/and shifting for curves
        with one bend, e.g. 15
    the payof of … in terms of more
        valid probability statements,
        e.g. 412
    to achieve:
        linearity 83, 300, 547, 916, 1001,
            1098, 1153, 1320, 1383
        normality, e.g. 547, 716, 916,
            943, 955, 956, 1052, 1320
    to linearize data 83, 300, 547, 916,
        1001, 1098, 1153, 1320, 1383
    when are … worthwile? e.g. 547
Transforming for symmetry, e.g. 547
Transition frequency matrix and
    dyadic interactions 1365
Translation versus standardization
    1428
Transport: Logistik 293, 333
Travelling salesman problem 293,
    333, 867, 1022
Treatment versus control in
    randomized block designs,
    extended tables of the
    distributions of rank statistics
    for 980
Tree:
    classification 132
    diagrams, the assignment of ob-

jects by binary, nominal and ordinal variables 132

networks, e.g. 1333

-structured methods 132

Trendanalyse: siehe auch Zeitreihen-Analyse

Trend:

and seasonal adjustment: seven graphical displays for judging the adequacy of the seasonal adjustment of a series and for understanding the behavior of the trend, seasonal and irregular components 202

estimation: split middle method, e.g. 527

Trends:

cyclical, in incidence data 1126

in ordered contingency tables, exact tests for 1027

in Poisson- oder binomialverteilten Grundgesamtheiten 58

stochastic and deterministic, and seasonality, e.g. 1019

Trend surface analysis, e.g. 1333

Trials, clinical: see clinical trials

Triangular:

coordinates, e.g. 224

form, orthogonal reduction to upper, e.g. 661, 824, 1205

reduction and linear models, e.g. 661, 824, 1205

Triangulation of large real world input-output matrices, e.g. 456

Triefficiency 547

Trigamma distribution 632

Trilinear models 417

Trimean and hinge, e.g. 547, 1319, 1320

Trimmed:

means 515, 547, 774, 1248

standard deviation 515

Trimming und Winsorization 287, 1358

True-score analysis of covariance 585

Truncated normal distribution 195

Trustworthiness of computer hardware and networks, e.g. 46

TSPACK 181

T square distribution, e.g. 22, 634, 800, 840, 1089

t-statistic, tables of the Bonferroni 38

t table for simultaneous control on g contrasts 391

t test:

and the Wilcoxon-Mann-Whitney test 533

exact p values 744

Tuberculosis, age-period-cohort analysis of 851

Tukey's:

quick test: new tables 394

test for nonadditivity, e.g. 240, 674, 732

Tumor growth curve analysis, e.g. 703

Tumorigenicity experiments, analysis of rodent 198

Tumorigenitätsstudien, z. B. 1353

Tumorkranke: Basisdokumentation 1356

Tumormarker in Diagnostik und Prognostik 2, 986, 1355

Tumor markers, diagnostic sensitivity, diagnostic specificity and predictive value of the determination of, e.g. 1355

Tumornachsorge mit Petrinetzen 692

t value approximations 400, 1290

t variables, range of independent, a table of percentage points of the 1397

t-Verteilung:

Approximationen 400, 1290

nichtzentrale 541, 633, 1005, 1028

TWAICE, e.g. 1163

Twicing, in smoothing, e.g. 871

Two-:

and multidimensional data, studying: graphical methods, e.g. 183

and three-dimensional tables: see also contingency tables and homogeneity of distributions

t-Verteilung:
Two-:
  partitioning of chi-square 103,
    263, 352, 793, 836, 1100,
    1153, 1334
  and three-way displays of one-
    way and two-way data 1359
by-two:
  and larger tables of frequen-
    cies: see contingency
    table analysis
  table chi-square test:
    see fourfold table analysis
    with (n-1) substituted for n
      in the numerator 1111,
      1151–1153, 1335
period crossover design, e.g.
  160
phase regression and change
  point problem 1212
sample:
  location and scale problems,
    e.g. 1090
  procedures in nonparametric
    statistics: graphical
    methods 363
  test of equal gamma distribu-
    tion scale parameters 1222
stage:
  least squares, e.g. 1420
  procedures for means, e.g.
    406
way:
  analysis by medians 547, 871,
    916, 1320, 1339
  classifications in the analysis
    of variance:
    unbalanced, e.g. 681, 732,
      780, 789, 1001, 1043, 1238,
      1259
    with equal or unequal num-
      bers and proportions 1238
data, unbalanced, fitting
  additive models to 1142
displays: graphical displays of
  two-way and multivariate
  data 1317, 1319, 1359, 1366

fit:
  plotting, e.g. 1319
  unbalanced data additive
    models, e.g. 1142
frequency tables, tests of: see
  contingency table analysis
PLUS analyses 547, 916, 1320,
  1339
tables:
  analysis by medians 547, 871,
    916, 1320, 1339
  median polish 547, 871, 916,
    1320, 1339
  resistant analysis of: see
    median polish
  response surface model
    diagnosis in 127
Typ einer Wechselwirkung: siehe
  Wechselwirkung
Type III errors, multiple comparisons
  and 111

Übereinstimmung von Rangreihen
  433, 434
Überlebenszeiten:
  siehe auch survival data
  Analyse mit Martingalmethoden
    1348
  Logranktest 21, 694, 893, 1291
  und ihre Analyse 134, 155, 173,
    184, 193, 234, 242, 332, 437,
    625, 649, 694, 761, 770, 864,
    888, 893, 950, 952, 974, 994,
    1060, 1198, 1226, 1291
Ulam's problem, e.g. 433
Umfang von Stichproben: siehe
  Stichprobenumfang
Umladeprobleme 293, 1052
Umwelt, physische und soziale 1079,
  1255
Unähnlichkeit und ihre Messung,
  z.B. 435
Unavailability, confidence limits for,
  e.g. 99

Unbalanced:
  ANOVA 780, 789, 1043
  data:
    analysis of 789, 1142
    analysis of variance, e.g. 780,
      789, 1043, 1142, 1259
    two-way data, fitting additive
      models to, e.g. 1142
Uncertainty:
  hunting out the real 640, 914
  judgement under: heuristics and
    biases 640
Uniform:
  and non-uniform random
    variables, generating, e.g. 129,
    661, 908
  log-odds model, e.g. 428
United Kingdom:
  statistical sources and techniques
    1106
  statistics in the, 1939–1945 35
Unsicherheit von Aussagen vor
  Gericht, z.B. 76
Unternehmen und Arbeitsstätten:
  Daten 1255
Untersuchungen: Einführung in ihre
  Planung, z.B. 1151
Unwahrscheinliches: Rätsel unserer
  Welt, z.B. 543
Urlaubs- und Erholungsreisen: Da-
  ten 1079, 1255
Ursache und Wirkung: siehe causal
  inference, causation und Kausa-
  lität
USA:
  summary of statistics on the social,
    political and economic status
    of government and business
    671
  Data on industrial, business,
    social, educational, financial
    and other topics 1372
  U.S. presidential primary campaign,
    model of the 128
U test 214, 285, 533, 793, 836, 1151,
  1152, 1336
  and mean ridit, e.g. 611

approximations 1336
as nonparametric analogue and
  extension of t test 533
für rechts zensierte Daten 834
Utility:
  measurement, e.g. 48, 324
  theory, e.g. 48, 87, 398, 654, 1119

**Vague concepts, the role of ... in data
  analysis** 915
Validating new methods in analytical
  chemistry 172, 289, 733, 1387
Validity of inference making in quasi-
  experimentation 220, 1404
VARCOV 703
Variabilität von Labordaten, z.B. 605
Variability, measures of, e.g. 739
Variable generation in survey and
  opinion research, e.g. 1243
Variables:
  dummy: see dummy variables
  indicator: see indicator variables
  lurking 121, 635
  or factors in epidemiologic studies,
    kinds of, e.g. 21, 682
  relationships between 274, 502,
    547, 658, 916, 1319, 1320, 1323
Variance, e.g. 658
  analysis of: see analysis of
    variance
  components 538, 628, 661, 716,
    732, 780, 1156, 1157, 1253,
    1390
  a bibliography 1156, 1157
  -covariance matrices: see
    covariance matrices
  estimation: methods, examples
    and programs 1419
  generalized, e.g. 634
  multivariate analysis of 12, 482,
    661, 688, 716, 780, 798, 835,
    840, 855, 1209, 1284
  of a normal distribution: confi-
    dence interval 802
  -reduction techniques, e.g. 129,
    760, 908, 1143

Variances:
  equality:
    of… of correlated normal
      variables 512
    of… of correlated variates with
      incomplete data 95
  homogeneity of:
    a set of… against ordered alter-
      natives, tests for the 387
    see also Bartlett's test
  nonparametric test for 214, 215,
    315, 782
  sample, exact statistical tolerance
    limits for 1306
  the assumption of equal, e.g. 126
  unequal: multiple comparisons of
    means 102, 203, 309, 314, 393,
    698, 1116, 1288, 1397
Varianzanalyse 83, 93, 285, 288, 412,
    517, 587, 658, 716, 732, 895, 911,
    955, 1001, 1093, 1238, 1259,
    1390, 1408
  siehe auch ANOVA
  bei heterogenen Varianzen 102,
    277, 305, 698, 755, 1288, 1397
  clustering methods for grouping
    means 170
  current controversies 1259
  feste Effekte 362, 412, 587
  geordnete Mittelwerte 938, 941
  graphische Darstellung von Mittel-
    werten 23
  heterogene Varianzen 102, 277,
    305, 698, 755, 1288, 1397
  mit Prozentzahlen, z.B. 799
  multivariate 12, 285, 482, 587, 634,
    661, 688, 716, 780, 798, 835,
    840, 855, 900, 972, 1209, 1284,
    1323
  nichtorthogonale 239, 780, 900
  parametrische Alternativen zur 203
  Power 525, 944, 1408
  Stichprobenumfänge 944, 1408
  unbekannte und ungleiche Varian-
    zen 102, 277, 305, 755, 1288,
    1397
  Zweiwegklassifizierung bei un-
    gleichen Besetzungszahlen
    780, 789
Varianzgleichheit: Einstichprobentest auf 802
Varianzheterogenität in der Varianzanalyse 102, 277, 305, 755, 1288,
  1397
Varianzhomogenität:
  F max-test 652
  korrelierter normalverteilter
    Zufallsvariablen 512
  korrelierter Variabler bei un-
    vollständigen Daten 95
  Test gegen geordnete Alternativen
    387
Varianzkomponenten: 587, 661, 716,
  732, 780, 1156, 1157, 1390
  in der biometrischen Genetik, z.B.
    852
Variates, correlated, with incomplete
  data, testing equality of
  variances of 95
Variation:
  across studies: cumulation and in-
    tegration of extracted infor-
    mation, e.g. 414, 588, 1132,
    1270
  indices of nominal data, e.g. 1110
Variationskoeffizient:
  bekannt: Schätz- und Testverfah-
    ren 267
  Vertrauensbereich 1370
Variations of analytical results in
  clinical medicine 441
Vector:
  and matrix operations for multi-
    variate analysis, e.g. 54, 443,
    445, 634, 1207
  geometry, e.g. 1420
  mean, inferences about a, e.g. 54,
    634
Vectors, random, basic properties of,
  e.g. 840
Veränderungsmessung 413, 1047
  bei Einzelfallbetrachtung 117, 413,
    527, 571, 1047, 1348
Verallgemeinerte lineare Modelle 6,

31, 288, 346, 520, 564, 573, 587, 674, 732, 814, 864, 922, 936, 1122
Verbrauch: Ermittlung des privaten Verbrauchs, ein Beitrag zur Fehlertheorie 1102
Verbrauchsstatistik 1079
Verbundene Stichproben: verteilungsfreie Homogenitätstests und Symmetrie-Hypothesen 708
Verdienste, Einkommen 1079, 1255
Verfahrensbibliothek 1096
Vergleichbarkeit klinischer Studien, z. B. 1348
Verfügbarkeit:
    eines Computers, z. B. 46, 283
    technischer Systeme, z. B. 46, 524, 847, 949, 952, 999, 1130, 1168, 1309
Vergleich:
    geordneter Mittelwerte bei Varianzhomogenität 734, 938
    mehrerer Behandlungen mit einer Kontrolle 68, 369, 787
    mehrerer Meßmethoden, z. B. 172
    mehrerer Mittelwerte bei ungleichen Varianzen 102, 277, 305, 1288, 1397
    mehrerer unabhängiger Stichproben gepaarter Beobachtungen 795
    von Überlebenszeiten 173, 184, 234, 242, 332, 625, 649, 694, 761, 770, 893, 952, 974, 1060, 1198, 1291
    von Verlaufskurven 703, 709, 772, 1041
Verhältnisschätzungen, z. B. 779
Verkehr: Daten 584, 1255
Verkehrsunfälle, ein mathematisches Modell zur Beschreibung von, z. B. 424
Verlaufskurvenanalyse, statistische Überlegungen zur: Regressionsmodelle mit zufälligen Koeffizienten, zweistufige Schätzverfahren 627

Verlaufskurven:
    Diskrimination und Klassifikation von 458
    nichtparametrische Regression für die Analyse von, z. B. 919
    Schätzung einer mittleren 241, 1041
    Vergleich von 703, 709, 772, 1041, 1432
    Prüfung der Symmetrie korrelierter Variablen 94
Verpackung 1052
Versicherungsmathematik 711
Versicherungsstatistik 64, 575, 758, 1300
Versuchsfehler: Nichtnormalität 235
Versuchsplanung 33, 98, 121, 123, 251, 270, 348, 413, 474, 477, 517, 525, 538, 578, 579, 587, 658, 674, 896, 955, 1036, 1076, 1093, 1096, 1107, 1122, 1246, 1257
Verteidigung: Daten 584
Verteilte Datenbanksysteme: siehe distributed data base systems
Verteilung, siehe auch distribution
Verteilungen:
    ansteckende 298
    multivariate: siehe multivariate distributions
    statistische 504, 541, 542, 631, 658, 1028, 1030, 1320
Verteilungsfreie:
    Methoden 214, 312, 363, 369, 703, 708, 709, 710, 719, 770, 793, 836, 893, 1024, 1072, 1090, 1225, 1426
    siehe auch nonparametric methods
    Tests, adaptive 157, 559, 879
    Vertrauensgrenzen für Quantile stetiger Verteilungen, z. B. 1426
Verteilungsfunktion, empirische, z. B. 643, 658, 1024, 1042
Verteilungspolygon und empirische Verteilungsfunktion 1042

Verteilungsunabhängige Methoden
214, 312, 363, 369, 703, 770, 793,
836, 893, 1024, 1072, 1090, 1225
Verteilungsvoraussetzungen, graphische Prüfung von, z. B. 183
Vertrauensbereich: siehe auch
confidence interval
Vertrauensbereiche:
(einseitige) Binomial- und
Poissonverteilung, mit
verbesserter Approximation
anhand der Normalverteilung
486
Einführung und Beispiele
1151-1153
Theorie 643, 658, 1090
Vertrauensgrenzen für Quantile stetiger Verteilungen, verteilungsfreie, z. B. 1426
VGR, Volkswirtschaftliche Gesamtrechnungen 20, 268, 1171, 1255
Vierfelderkorrelation 210, 263, 284,
367, 659, 739, 793, 798, 836, 1122
Vierfeldertest: siehe auch fourfold
tables
Vierfeldertests 103, 137, 227, 352, 367,
470, 643, 659, 793, 836, 884, 984,
1002-1004, 1056, 1111,
1151-1153, 1334, 1335, 1337
Visual display of quantitative information: see graphical displays
and displaying multivariate data
Vital statistics 671, 1106, 1255
Volks- und Berufszählung 161
Volkswirtschaftliche Gesamtrechnungen 20, 268, 1171, 1255
Vorausschätzung der Privathaushalte
803
Vorhersageverfahren nach Holt und
Winters sowie nach Box und
Jenkins, z. B. 4, 122, 185, 186
Vorsorgeuntersuchungen 419
Vorzeichentest 709, 766, 793, 836,
1152, 1153
Vote-counting procedures for integrating studies, dangers of using,
e. g. 414
Voting weights as power proxies 128

Wachstum in der Pflanzenzucht 666
Wachstumskurven 21, 44, 178, 300,
422, 566, 690, 703, 1065
Wachstumsprozesse: ihre mathematische Formulierung 966, 1065,
1173
Wachstumsrate, durchschnittliche in
der Wirtschaft 20, 268, 1171
Wages (statistics) 1106
Wahlanalyse, z. B. 563
Wahlbeteiligung 1079, 1255
Wahrscheinlichkeiten: Berechnung
anhand von Rekursionsformeln,
z. B. 904
Wahrscheinlichkeitsbegriff, z. B. 48,
768, 816
Wahrscheinlichkeitsdichten, z. B. 541
Wahrscheinlichkeitsfunktionen, z. B.
542
Wahrscheinlichkeitsrechnung 19,
118, 304, 454, 643, 768, 816, 823,
1093, 1152, 1153
Historisches: siehe history of statistics
Wahrscheinlichkeit, subjektive 1120
Waiting time:
distributions, e. g. 50, 874
onset distribution, e. g. 332
Waiting times:
in a Poisson process, e. g. 643, 874
tables 1014
Waksberg method, random digit
dialing sampling 497
Wald's distribution 372
Waller-Duncan k ratio test for
making pairwise comparisons
among treatment means, e. g.
1001
Walsh averages, e. g. 1072
Wanderungsbewegungen 37, 161,
340, 455, 591, 818, 1148, 1255,
1281
Wareneingangskontrolle 211
Warenkorb und Indexzahlen 349
Warfare models 581
Waring:
distribution, e. g. 1430
expansion, e. g. 1430

Warner's randomized response techniques, e.g. 311, 329, 820, 1260

Warning systems, optimal: a game theoretic model for the optimal use of information with random error and with error under the control of the other player 128

Wartbarkeit 265, 883

Warteschlangen 339, 347, 424, 695

Watson statistics: tests of symmetry 534

WBS, Wissensbasierte Systeme, z. B. 1108

Weak association between an exposure and a disease 1429

Wearout and component reliability, e.g. 99

Wechselwirkungen 131, 232, 417, 525, 674, 716, 1122

in Kontingenztafeln, z. B. 1311

Wechselwirkung und Test auf Additivität der Faktorwirkungen 1122

Weibull:

and other continuous distributions, mixtures of 343

distribution 41, 209, 234, 343, 475, 504, 541, 631, 735, 761, 762, 770, 847, 921, 952, 1028, 1186, 1211, 1215

a nomogram for estimating the three parameters of the 1211

three-parameter: easy parameter estimation with the aid of charts and tables 209

distributions, comparison of two, e.g. 770

-Verteilung 41, 209, 234, 343, 475, 504, 541, 631, 735, 761, 762, 770, 847, 921, 952, 1028, 1211, 1215

Weighted:

distributions arising out of methods of ascertainment: what population does a sample represent? 35

least squares procedures 10, 11, 83, 300, 472, 798, 936, 1110, 1383

Weissberg outlier test in regression analysis 219, 1264, 1383

Welch's two-sample t test, e.g. 1152, 1153

W-estimator, e.g. 547

Widowhood, duration of, e.g. 194

Wiener filtering, e.g. 718

Wilcoxon-Mann-Whitney: distribution, a comparison of some approximations to the 1336

test: see U test

Wilcoxon matched-pair signed rank test, e.g. 285, 709, 836, 1072, 1153

Wilcoxon's rank-sum test: see U test

Wilcoxon-Test für Paardifferenzen, z. B. 709, 793, 836, 1072, 1152, 1153

Wilcoxon two sample test, e.g. 214, 285, 533, 793, 836, 1151, 1152, 1336

Wildlife sampling, e.g. 966, 1000, 1064, 1208, 1210, 1391

Wilks' lambda, e.g. 501, 634, 1284

Wilson's e, e.g. 1110

Window plot 23

Winsorization, e.g. 253, 287, 547, 1358

Winters' additive and multiplicative seasonal forecast procedures: software, e.g. 4

Winters-Holt: see Holt-Winters methods

Wirtschaft, Energie, Verkehr: Daten 584, 1255

Wirtschaftsstatistik, z. B. 20, 268, 526, 650, 726, 727, 1171, 1265

Wishart distribution and statistics based on the, e.g. 22, 31, 840, 1078

Wissensbanken und Knowledge Base Management Systems, z. B. 1108

Wissensbasierte Systeme (WBS) 26, 1108, 1163

Wissenschaftliche Methode, z. B. 123, 182, 220, 229, 291, 303, 423, 525, 754, 882, 1024, 1066, 1318, 1321–1323, 1361, 1404, 1411
Wissenserwerb und maschinelles Lernen, z. B. 1163
Within-site and cross-site analyses, e. g. 889
Witness and law, e. g. 299, 356, 959
Wölbung einer Verteilung, z. B. 630, 739, 1040, 1152, 1153
Wörterbuch der Statistik, z. B. 657
Work or random time sampling, e. g. 1395
World:
    Fertility Survey, reflections on the 35
    War II, statistical developments in: an international perspective 35
Worst object selection, e. g. 406
Writing:
    questions and designing question-naires 36, 125, 167, 609, 1137, 1182, 1200, 1276
    technical papers, e. g. 328, 379, 1263
W-test and K^2-test, e. g. 1052

Yates algorithm 121, 438, 445, 463
Yates's correction 470, 854, 1003, 1004, 1111, 1335, 1434
Years of life lost: rates and indices, e. g. 602
Young tableaux, e. g. 433
Yule-Walker equations, e. g. 4

Zahlungsbilanz, z. B. 20
Zeichentest: siehe Vorzeichentest
Zeitreihen: siehe auch time series
Zeitreihen-Analyse 4, 20, 24, 43, 93, 114, 122, 140, 169, 177, 185, 186, 202, 220, 230, 255, 274, 312, 413, 444, 616, 518, 523, 571, 626, 650, 658, 672, 673, 718, 720, 769, 777, 778, 859, 865, 906, 928, 942, 953, 958, 1019, 1041, 1080, 1085, 1121, 1171, 1184, 1203, 1265
    demographische Anwendungen, z. B. 626
    einfache und graphische Methoden, z. B. 20, 43, 202, 268, 880, 942, 1171
    neuere Entwicklungen, z. B. 230, 518, 958
    Übersichten 122, 185, 186, 230, 312, 518, 523, 658, 718, 720, 778, 957, 1184
    und multiple Regression, z. B. 1420
    und System-Analyse, z. B. 1019
Zeitreihen:
    in der Psychologie 413, 444, 574
    saisonale Schwankungen in öko-nomischen, Analyse 1265
    Spektralanalyse von, z. B. 122, 185, 186, 285, 718, 777, 1080, 1184, 1323
Zeitreihenverfahren, demogra-phische Anwendungen neuerer 626
Zeitreihen von Meßwerten 241, 1041, 1376
Zeitreihenzerlegung: siehe Zeitreihen-Analyse
Zensierte Überlebensdaten: stati-stische Verfahren 155, 1177, 1198
Zentraler Grenzwertsatz, z. B. 643, 768
    Nichtanwendbarkeit auf endliche Grundgesamtheiten 1057
Zero-order and first-order models, e. g. 160
Zinc and cadmium: distribution in blood 1278
Zonal polynomials, e. g. 1078
Zufälligkeit, z. B. 312
Zufallsauswahl 98, 621, 658, 727, 729, 1061, 1169, 1201, 1261, 1274, 1431, 1433

Zufallszahlen:
  Erzeugung und Anwendungen:
    siehe auch Simulation
  Erzeugen und Testen von 129, 130,
    308, 1085, 1117, 1143, 1190
  gleichverteilte, Generatoren für,
    z. B. 45, 308
  normalverteilte, z. B. 421, 1035
Zufall und Statistik, z. B. 675
Zufallsvariablen, z. B. 643, 768,
  769
Zusammenhangsanalysen medizinischer Daten, z. B. 334, 1384
Zuverlässigkeit technischer Systeme
  32, 99, 265, 420, 475, 524, 536,
  544, 552, 691, 699, 762, 847, 883,
  948, 949, 952, 974, 1130, 1165,
  1168, 1216, 1217, 1267, 1304,
  1308, 1309, 1327

Zuverlässigkeitsschaltbilder, z. B. 552
Zuverlässigkeit und Bias 172, 265,
  506, 1074, 1387
Zwei- und mehrdimensionale Daten,
  graphische Darstellung, von:
  siehe graphische Methoden
Zweiperiodenwechselversuch: siehe
  Cross-over design
Zweistichproben-Rangtests:
  auf Lageunterschiede, z. B. 214,
    793, 1072
  auf Lageunterschiede: siehe auch
    U-Test
Zweistichprobentests: Spezialfälle,
  z. B. 1222
Zweistichproben-T^2-Test nach
  Hotelling, z. B. 800, 1089
Zwischenauswertungen von Thera-
  piestudien, z. B. 692, 694

4. List of Abbreviations of Journal Titles/Abkürzungen einiger Zeitschriften-Titel

ASTA   Allgemeines Statistisches Archiv
AMEP   American Journal of Epidemiology
AMST   The American Statistician
APST   Applied Statistics
BEHA   Behaviormetrika
BIJL   Biometrical Journal (up to/bis 1976 Biometrische Zeitschrift)
BICS   Biometrics
BIKA   Biometrika
CHRO   Journal of Chronic Diseases
COMP   Computational Statistics and Data Analysis
CRYS   Crystallographic Statistics
CSTH   Communications in Statistics – Theory and Methods
CSSM   Communications in Statistics – Simulation and Computation
EDUC   Journal of Educational Statistics
EDVM   EDV in Medizin und Biologie
INTR   International Statistical Review (up to/bis 1971 Rev. Intern. Statist. Inst.)
JASA   Journal of the American Statistical Association
JQTE   Journal of Quality Technology
JRSA   Journal of the Royal Statistical Society Series A
JRSB   Journal of the Royal Statistical Society Series B
JSCS   Journal of Statistical Computation and Simulation
MIME   Methods of Information in Medicine
MEKA   Metrika
PYBE   Psychologische Beiträge
PYBU   Psychological Bulletin
PYKA   Psychometrika

RSAP   Revue de Statistique Appliquée
SANA   Sankhyā, The Indian Journal of Statistics, Series A
SANB   Sankhyā, The Indian Journal of Statistics, Series B
SMED   Statistics in Medicine
SSNL   Statistical Software Newsletter
STNE   Statistica Neerlandica
STAN   The Statistician
TECS   Technometrics
TRIA   Controlled Clinical Trials
ZGHY   Zeitschrift für die gesamte Hygiene und ihre Grenzge-
       biete

5. References/Literatur

For some journals I use a code given on page 130.
Für einige Zeitschriften verwende ich den auf Seite 130
gegebenen Code.

1. ABDEL-HAMID, A.R., J.A.BATHER and G.B.TRUSTRUM (1982): The secretary problem with an unknown number of candidates. J. Appl. Probab. **19**, 619–630

2. ABEL, U., J.BERGER und E.WEBER (1984): MARKERTEST – Statistische Verfahren und Programme zur Validierung biologischer Marker. EDVM **15**, 117–125

3. ABELS, H. und H.DEGEN (1981): Handbuch des statistischen Schaubilds. Konstruktion, Interpretation und Manipulation von graphischen Darstellungen. (Vlg. Neue Wirtschaftsbriefe; 312 S.) Herne/Berlin

4. ABRAHAM, B. and J.LEDOLTER (1983): Statistical Methods for Forecasting. (Wiley; pp.445) Chichester and New York

5. ACKERMANN, H. and K.ABT (1984): Designing the sample size for non-parametric, multivariate tolerance regions. BIJL **26**, 723–734

6. ADENA, M.A. and S.R. WILSON (1982): Generalised Linear Models in Epidemiological Research: Case-Control-Studies. (INTSTAT; pp.165) Forestville, NSW 2087, Australia

7. ADER, H.J., D.J.KUIK, E.OPPERDOES and B.F.SCHRIEVER (1985): The use of conversational packages in statistical computing. SSNL **11**, 106–116

8. AFIFI, A. and Virginia A.CLARK (1984): Computer-Aided Multivariate Analysis. (Lifetime Learning Publications; pp.458) Belmont, Calif. 94002

9. AGRESTI, A. (1981): Measures of nominal-ordinal association. JASA **76**, 524–529

10. AGRESTI, A. (1983): A survey of strategies for modeling cross-classifications having ordinal variables. JASA **78**, 184–198

11. AGRESTI, A. (1984): Analysis of Ordinal Categorical Data. (Wiley; pp.287) New York

12. AHRENS, H. und J. LÄUTER (1981): Mehrdimensionale Varianzanalyse. Hypothesenprüfung, Dimensionserniedrigung, Diskrimination bei multivariablen Beobachtungen. 2. erw. Aufl. (Akademie-Vlg.; 238 S.) Berlin

13. ALDRICH, J. H. and F. D. NELSON (1985): Linear Probability, Logit and Probit Models. (Sage Publ. Series 07-045; pp. 94) Beverly Hills and London

14. ALFERS, D. and H. DINGES (1984): A normal approximation for beta and gamma tail probabilities. Z. Wahrscheinlichkeitstheorie verw. Gebiete **65**, 399-420

15. ALLEN, D. M. and F. B. CADY (1982): Analyzing Experimental Data by Regression. (Lifetime Learning Publ.; pp. 394) Belmont, Calif.

16. ALLISON , P. D. (1980): Analysing collapsed contingency tables without actually collapsing. American Sociological Review **45**, 123-130

17. ANDERSEN, E. B. (1982): Latent structure analysis: a survey. Scand. J. Statist. **9**, 1-12

18. ANDERSON, J. A. (1984): Regression and ordered categorical variables. With discussion. JRSS B **46**, 1-30

19. ANDERSON, O., W. POPP, M. SCHAFFRANEK, D. STEINMETZ und H. STENGER (1976): Schätzen und Testen. Eine Einführung in die Wahrscheinlichkeitsrechnung und schließende Statistik. (Heidelberger TB, Bd. 177) (Springer; 385 S.) Berlin, Heidelberg, New York

20. ANDERSON, O., M. SCHAFFRANEK, H. STENGER und K. SZAMEITAT (1983): Bevölkerungs- und Wirtschaftsstatistik. Aufgaben, Probleme und beschreibende Methoden. (Heidelberger TB, Bd. 223) (Springer; 444 S.) Berlin, Heidelberg, New York

21. ANDERSON, Sharon, Ariane AUQUIER, W. W. HAUCK, D. OAKES, W. VANDAELE and H. I. WEISBERG (A. S. BRYK and J. KLEINMAN) (1980): Statistical Methods for Comparative Studies. Techniques for Bias Reduction. (Wiley; pp. 289) New York

22. ANDERSON, T. W. (1984): An Introduction to Multivariate Statistical Analysis. 2nd ed. (Wiley; pp. 675) New York

23. ANDREWS, H. P., R. D. SNEE and Margaret H. SARNER (1980): Graphical display of means. AMST **34**, 195-199

24. ANSCOMBE, F. J. (1981): Computing in Statistical Science through APL. (Springer; pp. 426) New York, Heidelberg

25. ANSCOMBE, F. J. (1982): How much to look at the data. Utilitas Mathematica **21A**, 23-28

26. APPELRATH, H.-J. (1985): Die Erweiterung von DB- und IR-Systemen zu wissensbasierten Systemen. Nachrichten für Dokumentation **36**, 13-21 [vgl. auch S. 2-12 und 33-37]

27. ARMENIAN, H. K. and A. M. LILIENFELD (1983): Incubation period of disease. Epidemiologic Reviews **5**, 1-15

28. ARMITAGE, P. (1985): The search for optimality in clinical trials. With discussion. INTR **53**, 15-36

29. ARMSTRONG, J. S. and E. J. LUSK (1983): Commentary on the Makridakis time series competition. The Accuracy of alternative extrapolation models: Analysis of a forecasting competition through open peer review. Journal of Forecasting **2**, 259–311

30. ARNOLD, B. C. (1983): Pareto Distributions. (International Co-operative Publ. House; pp. 326) Burtonsville, MD

31. ARNOLD, S. F. (1981): The Theory of Linear Models and Multivariate Analysis. (Wiley-Interscience; pp. 475) New York

32. ASCHER, H. and H. FEINGOLD (1984): Repairable Systems Reliability: Modeling, Inference, Misconceptions and Their Causes. (M. Dekker; pp. 232) New York

33. ATKINSON, A. C. (1982): Developments in the design of experiments. INTR **50**, 161–177

34. ATKINSON, A. C. (1985): Plots, Transformations, and Regression. An Introduction to Graphical Methods of Diagnostic Regression Analysis. (Oxford University Press; pp. 282) Oxford

35. ATKINSON, A. C. and S. E. FIENBERG (Eds.; 1985): A Celebration of Statistics. The ISI Centenary Volume. A Volume to Celebrate the Founding of the International Statistical Institute in 1985. (Springer; pp. 606), New York

36. BACKSTROM, Ch. H. and G. HURSH-CESAR (1981): Survey Research. 2nd ed. (Wiley; pp. 436) New York

37. BÄHR, J. (1983): Bevölkerungsgeographie. Verteilung und Dynamik der Bevölkerung in globaler, nationaler und regionaler Sicht. (UTB 1249) (E. Ulmer; 427 S.) Stuttgart

38. BAILEY, B. J. R. (1977): Tables of the Bonferroni *t*-statistic. JASA **72**, 469–478

39. BAILEY, B. J. R. (1981): Alternatives to Hastings' approximation to the inverse of the normal cumulative distribution function. APST **30**, 275–276

40. BAILEY, R. A. (1985): Restricted randomization versus blocking. INTR **53**, 171–182

41. BAIN, L. J. and M. ENGELHARDT (1983): A review of model selection procedures relevant to the Weibull distribution. CSTH **12**, 589–609

42. BAKER, F. B. (1981): Log-linear, logit-linear models: a didactic. EDUC **6**, 75–102

43. BAMBERG, G. und F. BAUR (1984): Statistik. 3. überarb. Aufl. (Oldenbourg; 334 S.) München (4. Aufl. 1985)

44. BARCIKOWSKI, R. S. and R. R. ROBEY (1984): Decisions in single group repeated measures analysis: statistical tests and three computer packages. AMST **38**, 148–150

45. BAREL, M. und B. JOBES (1983): Untersuchungen an Generatoren für gleichverteilte Zufallszahlen. Angewandte Informatik **25**, 404–409

46. BARLOW, R. E. and N. D. SINGPURWALLA (1985): Assessing the reliability of computer software and computer networks: an opportunity for partnership with computer scientists. AMST **39**, 88–94

134

47. BARNETT, V. (Ed.; 1981): Interpreting Multivariate Data. (Conf. Proc.; Sheffield, UK; March 1980) (Wiley; pp.374) New York

48. BARNETT, V. (1982): Comparative Statistical Inference. 2nd ed. (Wiley; pp.325) London and New York

49. BARNETT, V. and T. LEWIS (1978): Outliers in Statistical Data. (Wiley; pp.384) New York [2nd ed., pp.478, 1985]

50. BARTHOLOMEW, D.J. (1982): Stochastic Models for Social Processes. 3rd ed. (Wiley; pp.384) New York

51. BARTHOLOMEW, D.J. (1983): Some recent developments in social statistics. INTR **51**, 1–9

52. BASAR, T. and G.J. OLSDER (1982): Dynamic Noncooperative Game Theory. (Academic Press; pp.480) London

53. BASAWA, I.V. and B.L.S.P. RAO (1980): Statistical Inference for Stochastic Processes. (Academic Press; pp.438) London

54. BASILEVSKY, A. (1983): Applied Matrix Algebra in the Statistical Sciences. (North-Holland; pp.389) New York

55. BATHER, J.A. (1985): On the allocation of treatments in sequential medical trials. With discussion. INTR **53**, 1–13 and 25–36

56. BATSCHELET, E. (1980): Einführung in die Mathematik für Biologen. (Springer; 557 S.) Heidelberg

57. BATSCHELET, E. (1981): Circular Statistics in Biology. (Academic Press; pp.388) London

58. BAUER, P. and P. HACKL (1978): Inference on trends in several Poisson or binomial populations. BIJL **20**, 645–654

59. BAUER, P. and P. HACKL (1985): The application of Hunter's inequality in simultaneous testing. BIJL **27**, 25–38

60. BAUER, P. and M. SCHEMPER (1984): Tests of intraclass dependence for qualitative variables. EDVM **15**, 114–117

61. BAYER, R., K. ELHARDT, W. KIESSLING und D. KILLAR (1984): Verteilte Datenbanksysteme. Informatik-Spektrum **7**, 1–19

62. BAYNE, C.K., J.J. BEAUCHAMP, V.E. KANE and G.P. MCCABE (1983): Assessment of Fisher and logistic linear and quadratic discrimination models. COMP **1**, 257–273

63. BEAN, S.J. and C.P. TSOKOS (1980): Developments in nonparametric density estimation. INTR **48**, 267–287

64. BEARD, R.E., T. PENTIKÄINEN and E. PESONEN (1984): Risk Theory. The Stochastic Basis of Insurance. 3rd ed. (Chapman and Hall; pp.408) London

65. BEASLEY, J.D. and S.G. SPRINGER (1977): The percentage points of the normal distribution. APST **26**, 118–121

66. BEAUMONT, G.P. (1980): Intermediate Mathematical Statistics. (Chapman and Hall; pp.248) London

67. BECHHOFER, R.E. and C.W. DUNNETT (1982): Multiple comparisons for orthogonal contrasts: examples and tables. TECS **24**, 213–222

68. BECHHOFER, R. E. and A. C. TAMHANE (1983): Design of experiments for comparing treatments with a control: tables of optimal allocations of observations. TECS **25**, 87–95

69. BECKER, R. A. and J. M. CHAMBERS (1984): *S* An Interactive Environment for Data Analysis and Graphics. (Wadsworth; pp. 550) Belmont, Calif.

70. BECKER, R. A. and J. M. CHAMBERS (1984): Extending the *S* System. (Wadsworth; pp. 144) Belmont, Calif.

71. BECKMAN, R. J. and R. D. COOK (1983): Outlier . . . . *s*. With discussion and response. TECS **25**, 119–163

72. BECKMAN, R. J. and G. L. TIETJEN (1973): Upper 10% and 25% points of the maximum F ratio. BIKA **60**, 213–214

73. BEGUN, J. M. and K. R. GABRIEL (1981): Closure of the Newman-Keuls multiple comparisons procedure. JASA **76**, 241–245 [and **78**, (1983), 949–957]

74. BEIRLANT, J., E. J. DUDEWICZ and E. C. VAN DER MEULEN (1982): Complete statistical ranking of populations, with tables and applications. Journal of Computational and Applied Mathematics (Belgien) **8**, 187–201

75. BELSLEY, D. A., E. KUH and R. E. WELSCH (1980): Regression Diagnostics: Identifying Influential Data and Sources of Collinearity. (Wiley; pp. 292) New York

76. BENDER, R., Susanne RÖDER und A. NACK (1981): Tatsachenfeststellung vor Gericht. Bd. I und II. (Beck; 241 S. und 198 S.) München

77. BENNETT, B. M. (1978): On a test for equality of dependent correlation coefficients. Statistische Hefte **19**, 71–76

78. BENNETT, B. M. (1982): On generalized indizes of diagnostic efficiency. BIJL **24**, 59–62

79. BENNETT, B. M. (1983): Further results in indices for diagnostic screening II. BIJL **25**, 453–457

80. BENNETT, B. M. (1984): Unconditional tests of hypotheses concerning relative risk. BIJL **26**, 765–769

81. BENNINGHAUS, H. (1982): Deskriptive Statistik. 4. Aufl. (Teubner; 280 S.) Stuttgart

82. BERENSON, M. L. (1982): A comparison of several *k* sample tests for ordered alternatives in completely randomized designs. PYKA **47**, 265–280 and 535–539

83. BERENSON, M. L., D. M. LEVINE and M. GOLDSTEIN (1983): Intermediate Statistical Methods and Applications. A Computer Package Approach. (Prentice-Hall; pp. 579) Englewood Cliffs

84. BERG, G. G. and H. D. MAILLIE (Eds.; 1981): Measurement of Risks. (Plenum Press; pp. 560) New York

85. BERG, L. (1985): Numerical results on growth functions. BIJL **27**, 565–580

86. BERGER, J. und K. H. HÖHNE (Hrsg.; 1983): Methoden der Statistik und

Informatik in Epidemiologie und Diagnostik. (Mediz. Informatik und Statistik, Bd. 40) (Springer; 451 S.) Berlin, Heidelberg, New York

87. BERGER, J. O. (1980): Statistical Decision Theory, Foundations, Concepts and Methods. (Springer; pp. 425) New York, Heidelberg, Berlin [2nd ed., pp. 617, 1985]

88. BERGMAN, S. W. and J. C. GITTINS (1985): Statistical Methods for Pharmaceutical Research Planning. (M. Dekker; pp. 272) New York

89. BERLEKAMP, E. R., J. H. CONWAY and R. K. GUY (1982): Winning Ways. Vol. 1: Games in General, Vol. 2: Games in Particular. (Academic Press; pp. 472 and 480) New York

90. BERTIN, J. (1982): Graphische Darstellungen und die graphische Weiterverarbeitung der Information. Übers. und bearb. v. W. Scharfe (de Gruyter; 275 S.) Berlin und New York

91. BETH, T. (Ed.; 1983): Cryptography. (Lect. Notes in Comp. Sci., Vol. 149) (Springer; pp. 402) Berlin, Heidelberg, New York

92. BETH, T., P. HESS und K. WIRL (1983): Kryptographie. Eine Einführung in die Methoden und Verfahren der geheimen Nachrichtenübermittlung. (Teubner; 205 S.) Stuttgart

93. BEUTEL, P. und W. SCHUBÖ (1983): SPSS 9 – Statistik-Programm-System für die Sozialwissenschaften. Eine Beschreibung der Programmversionen 8 und 9 nach N. H. Nie und C. H. Hull. 4. Aufl. (G. Fischer; 323 S.) Stuttgart, New York

94. BHAPKAR, V. P. and A. P. GORE (1973): A distribution-free test for symmetry in hierarchical data. Journal of Multivariate Analysis **3**, 483–489

95. BHOJ, D. S. (1984): On testing equality of variances of correlated variates with incomplete data. BIKA **71**, 639–641

96. BIEFANG, S., W. KÖPCKE and M. A. SCHREIBER (1983): Manual for the Planning and Implementation of Therapeutic Studies. (Lecture Notes in Medical Informatics, Vol. 20) (Springer; pp. 100) New York, Heidelberg, Berlin

97. BILLARD, L. (Ed.; 1985): Computer Science and Statistics. (Proc. 16th Symp. on the Interface, Atlanta, GA, USA, March 1984) (North-Holland; pp. 296) New York and Amsterdam

98. BILLETER-FREY, E. P. und V. VLACH (1982): Grundlagen der statistischen Methodenlehre. (UTB 1163) (G. Fischer; 429 S.) Stuttgart

99. BILLINTON, R. and R. N. ALLAN (1983): Reliability Evaluation of Engineering Systems. Concepts and Techniques. (Plenum Publ. Corp.; pp. 349) New York

100. BIRCH, J. B. (1983): On the power of robust tests in analysis of covariance. CSSM **12**, 159–182

101. BIRNBAUM, I. (1981): An Introduction to Causal Analysis in Sociology. (Macmillan; pp. 167) London

102. BISHOP, T. A. (1979): Some results on simultaneous inference for analysis of variance with unequal variances. TECS **21**, 337–340

103. BISCHOP, Yvonne M. M., S. E. FIENBERG and P. W. HOLLAND (LIGHT, R. J. and F. MOSTELLER) (1978): Discrete Multivariate Analysis: Theory and Practice. (MIT Press; pp. 557) Cambridge, Mass.

104. BLASER, A. und P. PISTOR (Hrsg.; 1985): Datenbanksysteme für Büro, Technik und Wissenschaft. GI-Fachtagung, Karlsruhe 20.-22. März 1985. (Informatik-Fachberichte, Bd. 94) (Springer; 519 S.). Berlin, Heidelberg, New York, Tokyo

105. BLYTH, C. R. and H. A. STILL (1983): Binomial confidence intervals. JASA **78**, 108-116

106. BOARDMAN, T. J. (1982): The future of statistical computing on desktop computers. AMST **36**, 49-58

107. BOARDMAN, T. J. and M. C. BRYSON (1978): A review of some smoothing and forecasting techniques. JQTE **10**, 1-11

108. BOCK, H.-H. (1980): Clusteranalyse – Überblick und neuere Entwicklungen. OR Spektrum **1**, 211-232

109. BOCK, H.-H. (1984): Explorative Datenanalyse – eine Übersicht. ASTA **68**, 1-40

110. BOETTCHER, W., W. HERRLITZ, E. NÜNDEL und B. SWITALLA (1983): Sprache: Das Buch, das alles über Sprache sagt. (Westermann; 368 S.) Braunschweig

111. BOFINGER, Eve (1985): Multiple comparisons and type III errors. JASA **80**, 433-437

112. BOISSEL, J. P. and C. R. KLIMT (Eds.; 1979): Multi-Center Controlled Trials. (Inst. Nat. Santé Rech. Med.; pp. 274) Paris

113. BOIVIN, J.-F. and S. WACHOLDER (1985): Conditions for confounding of the risk ratio and of the odds ratio. AMEP **121**, 152-158

114. BOLLINGER, G., A. HERRMANN und V. MÖNTMANN (1982): BMDP. Statistikprogramme für die Bio-, Human- und Sozialwissenschaften; nach W. J. Dixon ... und J. D. Toporek. Eine Beschreibung der Programmversionen 77-81. (G. Fischer; 431 S.) Stuttgart

115. BONETT, D. G. and P. M. BENTLER (1983): Goodness-of-fit procedures for the evaluation and selection of log-linear models. PYBU **93**, 149-166

116. BOOMSMA, A. (1977): Comparing approximations of confidence intervals for the product-moment correlation coefficient. STNE **31**, 179-185

117. BORTZ, J. (1984): Lehrbuch der empirischen Forschung. Für Sozialwissenschaftler. (Springer; 649 S.) Berlin, Heidelberg, New York, Tokyo [2. Aufl., 898 S., 1985]

118. BOSCH, K. (1982): Elementare Einführung in die Wahrscheinlichkeitsrechnung. Mit 82 Beispielen und 73 Übungsaufgaben mit vollständigem Lösungsweg. 3. durchges. Aufl. (Vieweg; 192 S.) Braunschweig

119. BOWMAN, K. O. and L. R. SHENTON (1982): Properties of estimators for the gamma distribution. CSSM **11**, 377-519

120. BOX, G. E. P. (1982): Choice of response surface design and alphabetic optimality. Utilitas Mathematica **21B**, 11-55

121. Box, G. E. P., W. G. Hunter and J. S. Hunter (1978): Statistics for Experimenters. An Introduction to Design, Data Analysis, and Model Building. (Wiley; pp. 653) New York

122. Box, G. E. P. and G. M. Jenkins (1976): Time Series Analysis: forecasting and control. Rev. ed. (Holden-Day; pp. 575) San Francisco and London

123. Box, G. E. P., T. Leonard and C.-F. Wu (Eds.; 1983): Scientific Inference, Data Analysis and Robustness. Proc. Conf. Mathematics Res. Ctr. University of Wisconsin, Nov. 1981. (Academic Press; pp. 320) New York and London

124. Boyer, J. E., Jr., A. D. Palachek and W. R. Schucany (1983): An empirical study of related correlation coefficients. EDUC **8**, 75-86

125. Bradburn, N. M. and S. Sudman (1980): Improving Interview Method and Questionnaire Design. (Jossey-Bass Publ.; pp. 214) San Francisco

126. Bradley, J. V. (1984): The complexity of nonrobustness effects. Bulletin of the Psychonomic Society **22**, 250-253

127. Bradu, D. (1984): Response surface model diagnosis in two-way tables. CSTH **13**, 3059-3106

128. Brams, S. J., A. Schotter and G. Schwödiauer (Eds.; 1979): Applied Game Theory. (Physica-Vlg.; pp. 447) Würzburg, Wien

129. Bratley, P., B. L. Fox and L. E. Schrage (1983): A Guide to Simulation. (Springer; pp. 385) New York, Heidelberg, Berlin

130. Brauer, W. (Hrsg.; 1982): Simulationstechnik. (Springer; 544 S.) Berlin, Heidelberg, New York

131. Bredenkamp, J. (1982): Verfahren zur Ermittlung des Typs einer statistischen Wechselwirkung. PYBE **24**, 56-74 und 309

132. Breiman, L., J. H. Friedman, R. A. Olshen and C. J. Stone (1984): Classification and Regression Trees. (Wadsworth; pp. 358) Belmont, Calif.

133. Brennecke, R., E. Greiser, H. A. Paul und Elisabeth Schach (Hrsg.; 1981): Datenquellen für Sozialmedizin und Epidemiologie. (Mediz. Informatik und Statist., Bd. 29) (Springer; 277 S. und eine Klapptafel) Berlin, Heidelberg, New York

134. Breslow, N. E. (1979): Statistical methods for censored survival data. Environmental Health Perspectives **32**, 181-192

135. Breslow, N. E. (1982): Design and analysis of case-control studies. Ann. Rev. Public Health **3**, 29-54

136. Breslow, N. E. (1984): Elementary methods of cohort analysis. International Journal of Epidemiology **13**, 112-115

137. Breslow, N. E. and N. E. Day (1980): Statistical Methods in Cancer Research. Vol. 1: The Analysis of Case-Control Studies. (IARC Scientific Publ. No. 32, International Agency for Research on Cancer; pp. 338) Lyon

138. Breslow, N. E., J. H. Lubin, P. Marek and B. Langholz (1983): Multiplicative models and cohort analysis. JASA **78**, 1-12

139

139. BRESLOW, N. E. and B. E. STORER (1985): General relative risk functions for case-control studies. AMEP **122**, 149–162

140. BRILLINGER, D. R. (Ed.; 1984): The Collected Works of John W. Tukey. Vol. I + II: Time Series 1949/64, 1965/84. (Wadsworth; pp. 689 and 656) Belmont, Calif.

141. BRILLINGER, D. R. and P. R. KRISHNAIAH (Eds.; 1983): Time Series in the Frequency Domain. (Handbook of Statistics, Vol. 3) (North-Holland; pp. 482) Amsterdam and New York

142. BRITTAIN, E., J. SCHLESSELMAN and B. V. STADEL (1981): Cost of case-control studies. AMEP **114**, 234–243

143. BROEMELING, L. D. (1985): Bayesian Analysis of Linear Models. (M. Dekker; pp. 472) New York

144. BROWN, B. W., M. E. ROZELL and E. Y. WEBSTER (1983): SURVAN – A nonparametric survival analysis package. Computer Science and Statistics [Proc. 15th Symp. Interface, Houston TX; J. E. Gentle (Ed.), Publ.: North-Holland, Amsterdam/New York] 87–92

145. BROWN, C. C. (1983): The statistical comparison of relative survival rates. BICS **39**, 941–948

146. BROWN, C. C. and S. B. GREEN (1982): Additional power computations for designing comparative Poisson trials. AMEP **115**, 752–758

147. BROWN, M. L. (1982): Robust line estimation with errors in both variables. JASA **77**, 71–79

148. BROWN, P. J. (1982): Multivariate calibration. JRSB **44**, 287–321

149. BROWNIE, C., J.-P. HABICHT and D. S. ROBSON (1983): An estimation procedure for the contaminated normal distributions arising in clinical chemistry. JASA **78**, 228–237

150. BRUCKMANN, G. (Hrsg.; 1977): Langfristige Prognosen. Möglichkeiten und Methoden der Langfristprognostik komplexer Systeme. (Physica-Vlg.; 458 S.) Würzburg

151. BRUZZI, P., S. B. GREEN, D. P. BYAR, L. A. BRINTON and C. SCHAIRER (1985): Estimating the population attributable risk for multiple risk factors using case-control data. AMEP **122**, 904–914

152. BRYANT, J. L. and N. T. BRUVOLD (1980): Multiple comparison procedures in the analysis of covariance. JASA **75**, 874–880

153. BUCHANAN, B. G. (1982): New research on expert systems. Machine Intelligence **10**, 269–299

154. BUCK, W. (1980): Tests of significance for point-biserial rank correlation coefficients in the presence of ties. BIJL **22**, 153–158

155. BUDDE, M. und M. WARGENAU (1984): Analyse von Überlebensdaten in SAS und BMDP. SSNL **10**, 126–135

156. BÜHLER, W. u. a. (Eds.; 1983): Operations Research Proceedings 1982. (Springer; 606 S.) Berlin, Heidelberg, New York

157. BÜNING, H. (1983): Adaptive verteilungsfreie Tests. Statist. Hefte **24**, 47–67

158. Bukac, I. and H. Burstein (1980): Approximations of Student's t and chi-square percentage points. CSSM B **9**, 665–672

159. Bulmer, M. G. (1985): The Mathematical Theory of Quantitative Genetics. (Oxford Univ. Press; pp. 262) Oxford

160. Buncher, C. R. and J.-Y. Tsay (Eds.; 1981): Statistics in the Pharmaceutical Industry. (M. Dekker; pp. 465) New York and Basel

161. Bundesinstitut für Bevölkerungsforschung (Hrsg.; 1984): Demographische Fakten und Trends in der Bundesrepublik Deutschland. Eine Bestandsaufnahme anläßlich der Internationalen Bevölkerungskonferenz 1984 der Vereinten Nationen. Juli 1984. Zeitschrift für Bevölkerungswissenschaft **10**, 295–397

162. Bundesminister des Innern (Hrsg.; 1983): Die Krebssterblichkeit in der Bundesrepublik Deutschland 1970–1978. Bd. I–III. (Vlg. TÜV Rheinland; [jeweils unterschiedliche Paginierungen]) Köln

163. Bundesminister für Jugend, Familie und Gesundheit (Hrsg.; 1983): Daten des Gesundheitswesens. Ausgabe 1983. (Band 152) (W. Kohlhammer; 355 S.) Stuttgart, Berlin, Köln, Mainz

164. Buyse, M. (1983): Les differentes approches conduisant a l'analyse des correspondances. Biométrie-Praximétrie **23**, 1–26

165. Buzzi-Ferraris, G. and P. Forzatti (1983): A new sequential experimental design procedure for discriminating among rival models. Chemical Engineering Science **38**, 225–232

166. Campbell, H. (Overall Ed.; 1980): Manual of Mortality Analysis. (Reprint of the 1977 ed.) (W. H. O.; pp. 245) Geneva

167. Cannell, C. F. and L. Oksenberg (1983): New questionnaire design for reducing response errors. Bull. Int. Statist. Inst. **50**, 500–511

168. Caporal, P. M. and G. J. Hahn (1984): General software for statistical graphics. A survey. SSNL **10**, 3–13

169. Cappella, J. N. (1980): Structural Equation Modeling: An Introduction. Chapter 3 in P. R. Monge and J. N. Cappella (Eds.): Multivariate Techniques in Human Communication Research. (Academic Press; pp. 552) New York and London, pp. 57–109

170. Carmer, S. G. and W. T. Lin (1983): Type I error rates for divisive clustering methods for grouping means in analysis of variance. CSSM **12**, 451–466

171. Carpenter, T. E. (1984): Epidemiologic modeling using a microcomputer spreadsheet package. AMEP **120**, 943–951

172. Carter, R. L. (1981): Restricted maximum likelihood estimation of bias and reliability in the comparison of several measuring methods. BICS **37**, 733–741

173. Carter, W. H., Jr., G. L. Wampler and D. M. Stablein (1983): Regression Analysis of Survival Data in Cancer Chemotherapy. (M. Dekker; pp. 209) New York

174. Casella, G. (1983): Leverage and regression through the origin. AMST **37**, 147–152

175. Casella, G. (1985): An introduction to empirical Bayes data analysis. AMST **39**, 83–87

176. Cassel, C.-M., C.-E. Sarndal and J. H. Wretman (1977): Foundations of Inference in Survey Sampling. (Wiley; pp. 192) New York

177. Catalano, R. A., D. Dooley and R. Jackson (1983): Selecting a time-series strategy. PYBU **94**, 506–523

178. Causton, D. R. and J. C. Venus (1981): The Biometry of Plant Growth. (E. Arnold; pp. 307) London

179. Cellier, F. E. (Ed.; 1982): Progress in Modelling and Simulation. (Academic Press; pp. 466) New York

180. Chaffin, W. W. and W. K. Talley (1980): Individual stability in Delphi studies. Technological Forecasting and Social Change **16**, 67–73

181. Chaghaghi, F. S. (1985): Time Series Package (TSPACK). (Lecture Notes in Computer Science, Vol. 187) (Springer; pp. 305) Berlin, Heidelberg, New York, Tokyo

182. Chalmers, A. F. (1978): What is this thing called Science. An Assessment of the Nature and Status of Science and its Methods. (The Open Univ. Press; pp. 157) Milton Keynes, England

183. Chambers, J., W. Cleveland, B. Kleiner and P. Tukey (1983): Graphical Methods for Data Analysis. (Wadsworth; pp. 395) Belmont, Calif.

184. Chan, M. Y., A. C. Cohen and Betty Jones Whitten (1983): The standardized inverse Gaussian distribution tables of the cumulative probability function. CSSM **12**, 423–442

185. Chatfield, C. (1982): Analyse von Zeitreihen. Eine Einführung. Übers. aus d. Engl. (2nd ed. 1980) durch H. Grimm. (Teubner; 239 S.) Leipzig

186. Chatfield, C. (1984): The Analysis of Time Series. An Introduction. 3rd ed. (Chapman & Hall; pp. 265) London

187. Chaum, D. (Ed.; 1984): Advances in Cryptology. Proceedings of Crypto 83. (Plenum Publ. Corp.; pp. 408) New York

188. Chen, H. J. (1979): Percentage points of multivariate t distribution with zero correlations and their application. BIJL **21**, 347–359

189. Chen, H. J. and A. W. Montgomery (1975): A table for interval estimation of the largest mean of k normal populations. BIJL **17**, 411–414

190. Chen, H. J. and P. J. Tsai (1980): Optimal confidence interval for the largest mean in repeated measurements design. BIKA **67**, 119–126

191. Cheng, S. W. and J. C. Fu (1983): An Algorithm to obtain the critical values of the t, χ^2 and F distributions. Statistics and Probability Letters (Amsterdam) **1**, 223–227

192. Chervany, N. L., P. G. Benson and R. K. Iyer (1980): The planning stage in statistical reasoning. AMST **34**, 222–226

193. Chiang, C. L. (1980): An Introduction to Stochastic Processes and their Applications. (R. E. Krieger Publ. Co.; pp. 517) Huntington, N. Y.

194. Chiang, C. L. (1984): The Life Table and its Applications. (R. E. Krieger; pp. 316) Malabar, Florida

195. CHIU, W. K. and M. P. Y. LEUNG (1981): A graphical method for estimating the parameters of a truncated normal distribution. JQTE **13**, 42-45

196. CHURCH, B. M. (1982): Recent developments in the Rothamsted general survey program. Utilitas Mathematica **21B**, 341-353

197. CHYTIL, M. K. (1983): Data preprocessing and computational analysis. SSNL **9**, 3-16

198. CLAYSON, D. B., D. KREWSKI and I. MUNRO (Eds.; 1985): Toxicological Risk Assessment. Vol. I and II (CRC Press; pp. 272, 288) Boca Raton/Florida 33431

199. CLEARY, J. P. and H. LEVENBACH (1982): The Professional Forecaster: The Forecasting Process Through Data Analysis. (Lifetime Learning Publ.; pp. 402) Belmont, Calif.

200. CLEVELAND, W. S. and R. MCGILL (1984): Graphical perception: theory, experimentation, and application to the development of graphical methods. JASA **79**, 531-554 [see also AMST **38** (1984), 261-280 and **39** (1985), 238-239]

201. CLEVELAND, W. S. and R. MCGILL (1985): The many faces of a scatterplot. JASA **79**, 807-822

202. CLEVELAND, W. S. and Irma J. TERPENNING (1982): Graphical methods for seasonal adjustment. JASA **77**, 52-62

203. CLINCH, J. C. and H. J. KESELMAN (1982): Parametric alternatives to the analysis of variance. EDUC **7**, 207-214

204. CLOGG, C. C. (1982): Some models for the analysis of association in multiway cross-classifications having ordered categories. JASA **77**, 803-815

205. COCHRAN, W. G. (MOSES, L. E. and F. MOSTELLER, Eds.; 1983): Planning and Analysis of Observational Studies. (Wiley; pp. 145) New York

206. CODY, R. P. and J. K. SMITH (1985): Applied Statistics and the SAS Programming Language. (Elsevier Sci. Publ.; pp. 192) Amsterdam and New York

207. COHEN, A. C. and Betty Jones WHITTEN (1980): Estimation in the three-parameter lognormal distribution. JASA **75**, 399-404

208. COHEN, A. C. and Betty Jones WHITTEN (1981): Estimation of lognormal distributions. American Journal of Mathematical and Management Sciences **1**, 139-153

209. COHEN, A. C., Betty Jones WHITTEN and Yihua DING (1984): Modified moment estimation for the three-parameter Weibull distribution. JQTE **16**, 159-167

210. COHEN, J. and P. COHEN (1975): Applied Multiple Regression Analysis. Correlation Analysis for the Behavioral Sciences. (Wiley; pp. 490) New York

211. v. COLLANI, E. (1984): Optimale Wareneingangskontrolle. Das Minimax-Regret-Prinzip für Stichprobenpläne beim Ziehen ohne Zurücklegen (Teubner; 150 S.) Stuttgart

212. COLMAN, A. M. (1982): Game Theory and Experimental Games. The

Study of Strategic Interaction. (Pergamon Press; pp. 301) Oxford, New York, Frankfurt

213. CONNELL, F. A. and T. D. KOEPSELL (1985): Measures of gain in certainty from a diagnostic test. AMEP **121**, 744–753

214. CONOVER, W. J. (1980): Practical Nonparametric Statistics. 2nd ed. (Wiley; pp. 510) London

215. CONOVER, W. J. and R. L. IMAN (1978): Some exact tables for the squared ranks test. CSSM B **7**, 491–513

216. CONOVER, W. J. and R. L. IMAN (1980): The rank transformation as a method of discrimination with some examples. CSTH A **9**, 465–487

217. CONOVER, W. J. and R. L. IMAN (1981): Rank transformations as a bridge between parametric and nonparametric statistics. With comments and rejoinder. AMST **35**, 124–133

218. COOK, Nancy R. and J. H. WARE (1983): Design and analysis methods for longitudinal research. Ann. Rev. Public Health **4**, 1–23

219. COOK, R. D. and S. WEISBERG (1982): Residuals and Influence in Regression. (Chapman and Hall; pp. 230) London and New York

220. COOK, T. D. and D. T. CAMPBELL (1979): Quasi-Experimentation. Design and Analysis Issues for Field Settings. (Rand McNally College Publ. Comp.; pp. 405) Chicago

221. COOPER, J. C. B. (1983): Factor analysis: an overview. AMST **37**, 141–147

222. CORMACK, R. M. and J. K. ORD (Eds.; 1979): Spatial and Temporal Analysis in Ecology. (International Co-operative Publ. House; pp. 356) Burtonsville, MD

223. CORMACK, R. M., G. P. PATIL and D. S. ROBSON (Eds.; 1979): Sampling Biological Populations. (International Co-operative Publ. House; pp. 392) Burtonsville, MD

224. CORNELL, J. A. (1981): Experiments with Mixtures: Design, Models and the Analysis of Mixture Data. (Wiley; pp. 305) New York

225. CORNELL, R. G. (Ed.; 1984): Statistical Methods for Cancer Studies. (M. Dekker; pp. 496) New York

226. COUSTEAU, J.-Y. (Hrsg.; 1983): Bestandsaufnahme eines Planeten. Cousteau-Umweltlesebuch 1. (Übers. aus d. Amerik. von Elke Martin) (Klett-Cotta; 199 S.) Stuttgart (insbes. „Die Welt in Zahlen", S. 89/183)

227. COX, D. R. (1970): The Analysis of Binary Data. (Methuen; pp. 142) London

228. COX, D. R. (1972): Regression models and life tables. With discussion. JRSB **34**, 187–220

229. COX, D. R. (1981): Theory and general principle in statistics. JRSA **144**, 289–297

230. COX, D. R. (1981): Statistical analysis of time series: some recent developments. Scandinavian Journal of Statistics **8**, 93–115

231. COX, D. R. (1982): Statistical significance tests. Brit. J. clin. Pharmac. **14**, 325–331

232. Cox, D. R. (1984): Interaction. With discussion. INTR **52**, 1–31

233. Cox, D. R. and Valerie Isham (1980): Point Processes. (Chapman and Hall; pp. 188) London and New York

234. Cox, D. R. and D. O. Oakes (1984): Analysis of Survival Data. (Chapman and Hall; pp. 201) London and New York

235. Cox, D. R. and E. J. Snell (1968): A general definition of residuals. With discussion. JRSB **30**, 248–275

236. Cox, D. R. and E. J. Snell (1981): Applied Statistics. Principles and Examples. (Chapman and Hall; pp. 189) London and New York

237. Cox, D. R. and E. Spjøtvoll (1982): On partitioning means into groups. Scand. J. Statist. **9**, 147–152

238. Coxon, A. P. M. (1982): The User's Guide to Multidimensional Scaling. With Special Reference to the MDS(X) Library of Computer Programs. (Heinemann Educat. Books; pp. 271) London

239. Cramer, E. M. and M. Appelbaum (1980): Nonorthogonal analysis of variance – once again. PYBU **87**, 51–57 [and **93**, (1983), 609]

240. Cressie, N. A. C. (1978): Removing nonadditivity from two-way tables with one observation per cell. BICS **34**, 505–513

241. Crowder, M. J. (1980): Proportional linear models. APST **29**, 299–303

242. Crowley, J. and R. A. Johnson (Eds.; 1982): Survival Analysis. Proceedings of the Special Topics Meeting sponsored by the Institute of Mathematical Statistics, Oct. 26–28, 1981, Columbus, Ohio (Institute of Mathematical Statistics; pp. 301) Hayward, California

243. Cummins, J. D., B. D. Smith, R. N. Vance and J. L. VanDerhei (1982): Risk Classification in Life Insurance. (Kluwer; pp. 320) Boston

244. Currie, L. A. (1968): Limits for qualitative detection and quantitative determination. Application to radiochemistry. Analytical Chemistry **40**, 586–593

245. Cyr, J. L. and E. B. Manoukian (1982): Approximate critical values with error bounds for Bartlett's test of homogeneity of variances for unequal sample sizes. CSTH **11**, 1671–1680

246. Däumler, K.-D. (1983): Finanzmathematisches Tabellenwerk für Praktiker und Studierende. 2. Aufl. (Vlg. Neue Wirtschafts-Briefe; 143 S.) Herne und Berlin

247. Dajani, J., M. Sincoff and W. K. Talley (1979): Stability and agreement criteria for the termination of Delphi studies. Technological Forecasting and Social Change **13**, 83–90

248. Dalenius, T. (1982): A sample of ideas for research and development in the theory and methods of sample surveys. Utilitas Mathematica **21A**, 59–74

249. Daniel, C. T. and F. S. Wood (1980): Fitting Equations to Data. Computer Analysis of Multifactor Data. 2nd ed. (Wiley; pp. 458) New York

250. Dannehl, K. und M. P. Baur (1983): Diagnostische und prognostische Tests – Anforderungen aus biostatistischer Sicht. Gynäkologe **16**, 177–187

251. Das, M. N. and N. C. Giri (1980): Design and Analysis of Experiments. (Wiley; pp. 308) New York

252. David, H. (1982): The Computer Package STATCAT. (North-Holland; pp. 780) New York and Amsterdam

253. David, H. A. (1981): Order Statistics. 2nd ed. (Wiley-Interscience; pp. 360) New York

254. Davidson, R. A. and P. H. Farquhar (1976): A bibliography on the method of paired comparisons. BICS **32**, 241-252

255. Davis, D. K. and J.-W. Lee (1980): Time-Series Analysis Models for Communication Research. Chapter 14 in P. R. Monge and J. N. Cappella (Eds.): Multivariate Techniques in Human Communication Research. (Academic Press; pp. 552) New York and London, pp. 429-454

256. Davis, R. (1982): Expert Systems: Where Are We? And Where Do We Going From Here? (A. I. Memo No. 665) (Massachusetts Institute of Technology; pp. 38) Cambridge, Mass.

257. Davison, M. L. (1983): Multidimensional Scaling. (Wiley; pp. 242) New York

258. Day, N. E., D. P. Byar and S. B. Green (1980): Overadjustment in case control studies. AMEP **112**, 696-706

259. Deaton, M. L. (1983): Estimation and hypothesis testing in regression in the presence of nonhomogeneous error variances. CSSM **12**, 45-66

260. Decouflé, P., T. L. Thomas and L. W. Pickle (1980): Comparison of the proportionate mortality ratio and standardized mortality ratio risk measures. AMEP **111**, 263-269

261. DeGroot, M. H. and S. E. Fienberg (1983): The comparison and evaluation of forecasters. STAN **32**, 12-22

262. Deichsel, G. and H. J. Trampisch (1985): Clusteranalyse und Diskriminanzanalyse. (G. Fischer, 135 S.) Stuttgart und New York

263. Delucchi, K. (1983): The use and misuse of chi-square: Lewis and Burke revisited. PYBU **94**, 166-176 (and **95**, 133)

264. Dempster, A. P., M. R. Selwyn, C. M. Patel and A. J. Roth (1984): Statistical and computational aspects of mixed model analysis. APST **33**, 203-214

265. DePriest, D. J. and R. L. Launer (Eds.; 1983): Reliability in the Acquisitions Process. (M. Dekker; pp. 216) New York

266. Deutler, T. (1976/1977): Die Bestimmung optimaler Schichtungen – ein Verfahrensvergleich. Jahrb. f. Nationalökonomie u. Statistik **191**, 153-173

267. Deutler, T. (1981): Schätz- und Testverfahren bei Normalverteilung mit bekanntem Variationskoeffizienten. (Springer; 195 S.) Heidelberg

268. Deutler, T., M. Schaffranek und D. Steinmetz (1984): Statistik-Übungen im wirtschaftswissenschaftlichen Grundstudium. (Heidelberger TB, Bd. 237) (Springer; 372 S.) Berlin, Heidelberg, New York

269. Diaconis, P. and B. Efron (1983): Computer intensive methods in sta-

tistics. Scientific American **248** (5), 96-109 and 138 (bzw. deutsche Version, Spektrum der Wissenschaften, Juli 1983, 96-109 und 183)

270. DIAMOND, W.J. (1981): Practical Experiment Designs for Engineers and Scientists. (Lifetime Learning Publications; pp.348) Belmont, Calif. 94002

271. DIDERICH, G.T. (1985): The Kalman filter from the perspective of Goldberger-Theil estimators. AMST **39**, 193-198

272. DIEHL, J.M. and H.-V. KOHR (1977): Durchführungsanleitungen für statistische Tests. (Beltz; 315 S.) Weinheim

273. DIELMAN, T. and R. PFAFFENBERGER (1982): LAV (Least Absolute Value) estimation in linear regression: a review. TIMS/Studies in the Management Sciences **19**, 31-52

274. DIELMAN, T.E. (1983): Pooled cross-sectional and time series data: a survey of current statistical methodology. AMST **37**, 111-122

275. DIELMAN, T.E. (1984): Least absolute value estimation in regression models: an annotated bibliography. CSTH **13**, 513-541

276. DIGBY, P.G.N. (1983): Approximating the tetrachoric correlation coefficient. BICS **39**, 753-757 [see also **40** (1984), 563]

277. DIJKSTRA, J.B. and P.S.P.J. WERTER (1981): Testing the equality of several means when the population variances are unequal. CSSM **10**, 557-569

278. DIJKSTRA, W. and J. VAN DER ZOUWEN (1982): Response Behaviour in the Survey-Interview. (Academic Press; pp.247) London

279. DIN Taschenbuch 22 (1978): Normen für Größen und Einheiten in Naturwissenschaft und Technik. AEF-Taschenbuch. 5. erweit. Aufl. (Beuth Vlg.; 403 S.) Berlin

280. DINGES, H. (1984): Auslastung von Telefonzellen. Mathematische Semesterberichte **31**, 77-94

281. DINKEL, R. (1983): Analyse und Prognose der Fruchtbarkeit am Beispiel der Bundesrepublik Deutschland. Zeitschrift für Bevölkerungswissenschaft **9**, 47-72

282. DINKEL, R. (1984): Sterblichkeit in Perioden- und Kohortenbetrachtung. Zeitschrift für Bevölkerungswissenschaft **10**, 477-500

283. DIRLEWANGER, W. (1983): Benutzer erwarten 99 Prozent Verfügbarkeit. Computerwoche 22.4., S.36/38; 29.4., S.34/35; 6.5., S.42/43; 15.7., S.30/31

284. DIVGI, D.R. (1979): Calculation of the tetrachoric correlation coefficient. PYKA **44**, 169-172

285. DIXON, W.J. (Ed.; 1981): BMDP Statistical Software 1981. (Univ. of Calif. Press; pp.727) Berkeley

286. DIXON, W.J. and F.J. MASSEY, Jr. (1983): Introduction to Statistical Analysis. 4th ed. (McGraw-Hill; pp.678) New York, Hamburg, London ...

287. DIXON, W.J. and K.K. YUEN (1974): Trimming and Winsorization: a review. Statistische Hefte **15**, 157-169

288. DOBSON, Annette J. (1983): An Introduction to Statistical Modelling. (Chapman and Hall; pp. 124) London

289. DOETSCH, K. (1983): Validating new methods. Clinical Chemistry **29**, 581

290. DOKSUM, K. A., Grete FENSTAD and R. AABERGE (1977): Plots and tests for symmetry. BIKA **64**, 473–487

291. DOLBY, G. R. (1982): The role of statistics in the methodology of the life sciences. BICS **38**, 1069–1083

292. DOLL, R. and R. PETO (1981): The causes of cancer: Quantitative estimates of avoidable risks of cancer in the United States today. J. Nat. Cancer Inst. **66**, 1191–1308

292. DOLL, R. and R. PETO (1981): The Causes of Cancer. Quantitative Estimates of Avoidable Risks of Cancer in the U. S. Today. (Oxford Univ. Press; pp. 144) New York

293. DOMSCHKE, W. (1981/82): Logistik: (1) Rundreisen und Touren. (2) Transport. Grundlagen, lineare Transport- und Umladeprobleme. (R. Oldenbourg; 176 S., 200 S.) München und Wien

294. DOMSCHKE, W. und A. DREXL (1984): Logistik: Standorte. (R. Oldenbourg; 221 S.) München und Wien

295. DONNER, A. (1982): The relative effectiveness of procedures commonly used in multiple regression analysis for dealing with missing values. AMST **36**, 378–381

296. DONNER, A. (1984): Approaches to sample size estimation in the design of clinical trials – a review. SMED **3**, 199–214

297. DOUGLAS, J. B. (1979): Analysis With Standard Contagious Distributions. (International Co-operative Publ. House; pp. 520) Burtonsville, MD

298. DOUGLAS, J. B. (1980): Analysis with Standard Contagious Distributions. (Internat. Co-op. Publ. House; pp. 520) Fairland, Maryland

299. DOWNTON, F. (1982): Legal probability and Statistics. With discussion. JRSA **145**, 395–402, 426–438

300. DRAPER, N. R. and H. SMITH (1981): Applied Regression Analysis. 2nd ed. (Wiley; pp. 709) New York

301. DRETZKE, B. J., J. R. LEVIN and R. C. SERLIN (1982): Testing for regression homogeneity under variance heterogeneity. PYBU **91**, 376–383

302. DREXL, A. (1982): Geschichtete Stichprobenverfahren. Optimierung der Schichtgrenzen und Stichprobenumfänge. (A. Hain, Meisenheim; 167 S.) Königstein/Ts.

303. DUBIN, R. (1978): Theory Building. 2nd rev. ed. (The Free Press/Macmillan; pp. 304) New York

304. DUDEWICZ, E. J. (1976): Introduction to Statistics and Probability. (Holt, Rinehart and Winston; pp. 512) New York and London

305. DUDEWICZ, E. J. and T. A. BISHOP (1981): Analysis of variance with unequal variances. JQTE **13**, 111–114

306. DUDEWICZ, E. J. and S. R. DALAL (1983): Multiple comparisons with a

control when variances are unknown and unequal. American Journal of Mathematical and Management Sciences **3**, 275–295

307. DUDEWICZ, E. J. and J. O. KOO (1982): The Complete Categorized Guide to Statistical Selection and Ranking Procedures. Vol. 6 (American Sciences Press; pp. 627) Columbus, Ohio

308. DUDEWICZ, E. J. and T. G. RALLEY (1981): The Handbook of Random Number Generation and Testing with TESTRAND Computer Code. (American Sciences Press; pp. 634) Columbus, Ohio

309. DUDEWICZ, E. J., J. S. RAMBERG and H. J. CHEN (1975): New tables for multiple comparisons with a control (unknown variances). BIJL **17**, 13–26

310. DÜRR, W. und K. KLEIBOHM (1983): Operations Research. Lineare Modelle und ihre Anwendungen. (Hanser; 332 S.) München und Wien

311. DUFFY, J. C. and Jennifer J. WATERTON (1984): Randomized response models for estimating the distribution function of a quantitative character. INTR **52**, 165–171

312. DUFOUR, J.-M., Y. LEPAGE and H. ZEIDAN (1982): Nonparametric testing for time series: a bibliography. Canad. J. Statist. **10**, 1–38

313. DUNNETT, C. W. (1980): Pairwise multiple comparisons in the homogeneous variance, unequal sample size case. JASA **75**, 789–795

314. DUNNETT, C. W. (1982): Robust multiple comparisons. CSTH **11**, 2611–2629

315. DURAN, B. S. (1976): A survey of nonparametric tests for scale. CSTH A **5**, 1287–1312

316. DYER, D. D. and J. P. KEATING (1980): On the determination of critical values for Bartlett's test. JASA **75**, 313–319

317. EAGLESON, G. K. (1983): A robust test for multiple comparisons of correlation coefficients. Australian Journal of Statistics **25**, 256–263

318. EBEL, S. (1983): Über den Vertrauensbereich kalibrierender Analysenverfahren. Teil 1: Lineare Kalibrierfunktion. Computer-Anwendung im Labor **1**, 55–61

319. EBEL, S., D. ALERT und U. SCHAEFER (1983): Über den Vertrauensbereich kalibrierender Analysenverfahren. Teil 2: Nichtlineare Kalibrierfunktionen bei Approximation durch Polynome. Computer-Anwendung im Labor **1**, 172–177

320. EBEL, S. und K. KAMM (1983): Statistische Definition der Bestimmungsgrenze. Fresenius Z. Anal. Chem. **316**, 382–385

321. EDGELL, S. E. and Sheila M. NOON (1984): Effect of violation of normality on the t test of the correlation coefficient. PYBU **95**, 576–583

322. EDGINGTON, E. S. (1980): Randomization Tests. (M. Dekker; pp. 287) New York

323. EDWARDS, D. and S. KREINER (1983): The analysis of contingency tables by graphical models. BIKA **70**, 553–565

324. EDWARDS, W. and J. R. NEWMAN (1982): Multiattribute Evaluation. (Sage Publicat. Series 07-026; pp. 95) Beverly Hills and London

325. EFRON, B. (1979): Bootstrap methods: another look at the jackknife. Annals of Statistics **7**, 1–26 [see also **10**, (1982), 212–225]

326. EFRON, B. (1982): The Jackknife, the Bootstrap and other Resampling Plans. (Society for Industrial and Applied Mathematics; pp. 92) Philadelphia

327. EFRON, B. and C. MORRIS (1977): Stein's paradox in statistics. Scientific American **236**, (5), 119–127 and 148

328. EHRENBERG, A. S. C. (1982): Writing technical papers or reports. AMST **36**, 326–329

329. EICHHORN, B. H. and L. S. HAYRE (1983): Scrambled randomized response methods for obtaining sensitive quantitative data. J. Statist. Planning Infer. **7**, 307–316

330. EICHHORN, C. (1984): Über die Anwendung verschiedener Diskriminationsverfahren. BIJL **26**, 193–197

331. ELANDT-JOHNSON, Regina C. (1971): Probability Models and Statistical Methods in Genetics. (Wiley; pp. 592) New York

332. ELANDT-JOHNSON, Regina C. and N. L. JOHNSON (1980): Survival Models and Data Analysis. (Wiley; pp. 457) New York

333. ELLINGER, T. (1984): Operations Research. Eine Einführung. (Springer; 269 S.) Berlin, Heidelberg, New York, Tokyo

334. ENKE, H. (1980): Zur Erfassung von Zusammenhangsstrukturen in mehrdimensionalen Kontingenztafeln. ZGHY **26**, 834–840

335. ENKE, H. (1981): Zur Prüfung spezieller Kontraste (Untertests) in 3-dimensionalen Kontingenztafeln zur Ausschöpfung bestimmter sachbezogener Informationen. ZGHY **27**, 387–394

336. ENKE, H. (1981): Zur Prüfung der Unabhängigkeit spezieller unvollständiger 2-dimensionaler Kontingenztafeln. ZGHY **27**, 796–799

337. EPPINK, T. W. A. (1984): Correspondence analysis versus principal component analysis for highly skewed distributed variables. Computational Statistics Quarterly (Vienna) **1**, 61–76

338. ERNSTER, Virginia (1983): The measurement of associations between environmental exposures and cancer. AMST **37**, 420–426

339. ESCHENBACH, W. (1984): Statistical inference for queuing models. Math. Operationsforsch. u. Statist., ser. statistics **15**, 451–462

340. ESENWEIN-ROTHE, Ingeborg (1982): Einführung in die Demographie. Bevölkerungsstruktur und Bevölkerungsprozeß aus der Sicht der Statistik. (F. Steiner; 400 S.) Wiesbaden

341. EUBANK, R. L. (1983): Approximate regression models and splines. (Techn. Rep. No. 180, Dept. Statist., Southern Methodist Univ.; pp. 56) Dallas, Texas

342. EVERITT, B. S. (1978): Graphical Techniques for Multivariate Data. (Heinemann; pp. 117) London

343. EVERITT, B. S. (1981): Finite Mixture Distributions. (Chapman and Hall; pp. 143) London

344. EVERITT, B.S. (1984): An Introduction to Latent Variable Models. (Chapman and Hall; pp. 107) London

345. FAGAN, T.J. (1975): Nomogram for Bayes' theorem. New England Journal of Medicine **293**, 257

346. FAHRMEIR, L. und A. HAMERLE (Hrsg.; 1984): Multivariate statistische Verfahren. (de Gruyter; 796 S.) Berlin

347. FAHRMEIR, L., H. KAUFMANN und F. OST (1981): Stochastische Prozesse. Eine Einführung in Theorie und Anwendungen. (Hanser; 366 S.) München, Wien

348. FEDERER, W.T. (1980/81): Some recent results in experimental design with a bibliography, I-III. INTR **48**, 357-368 and **49**, 95-109, 185-197

349. FERSCHL, F. (1980): Deskriptive Statistik. 2. verb. Aufl. (Physica-Vlg.; 308 S.) Würzburg und Wien

350. FIELLER, N.R.J. (1981): Looking at multivariate outliers from the consultant statistician's point of view. SSNL **7**, 38-46

351. FIENBERG, S.E. (1979): Graphical methods in statistics. AMST **33**, 165-178

352. FIENBERG, S.E. (1980): The Analysis of Cross-Classified Categorical Data. 2nd ed. (MIT Press; pp. 151) Cambridge, Mass.

353. FIENBERG, S.E. (1981): Recent advances in theory and methods for the analysis of categorical data: making the link to statistical practice. Bull. Int. Statist. Inst. **49**, 763-791

354. FIENBERG, S.E. and J.B. KADANE (1983): The presentation of Bayesian statistical analyses in legal proceedings. STAN **32**, 88-98

355. FIENBERG, S.E. and M.M. MEYER (1983): Loglinear models and categorical data analysis with psychometric and econometric applications. Journal of Econometrics **22**, 191-214

356. FIENBERG, S.E. and M.L. STRAF (1982): Statistical assessments as evidence. JRSA **145**, 410-421 [see also "law": 395-409 and 422-438]

357. FINCH, P.D. (1983): Do rates need standardizing for comparative purposes? BICS **39**, 275-279

358. FINK, E.L. (1980): Unobserved Variables within Structural Equation Models. Chapter 4 in P.R. Monge and J.N. Cappella (Eds.): Multivariate Techniques in Human Communication Research. (Academic Press; pp. 552) New York and London, pp. 111-141

359. FINNEY, D.J. (1977): Bioassay. In: Matthews, D.E. (Ed.): (Lecture Notes in Biomathematics, No. 18) Mathematics and the Life Sciences. (Springer; pp. 385) Berlin, Heidelberg, pp. 66-151

360. FINNEY, D.J. (1983): Biological assay: A microcosm of statistical practice. Bull. Int. Statist. Inst. **50**, 790-812

361. FISCHHOFF, B., P. SLOVIC and Sarah LICHTENSTEIN (1982): Lay foibles and expert fables in judgment about risk. AMST **36**, 240-255 (risk issue pp. 221-320)

362. FISHER, L. and J. McDONALD (1978): Fixed Effects Analysis of Variance. (Academic Press; pp. 192) New York

363. FISHER, N. I. (1983): Graphical methods in nonparametric statistics: a review and annotated bibliography. INTR **51**, 25-58

364. FISHER, N. I. and A. J. LEE (1983): A correlation coefficient for circular data. BIKA **70**, 327-332 (see also 333-341)

365. FISHER, R. A. and F. YATES (1982): Statistical Tables for Biological, Agricultural and Medical Research. 6. ed. (Longman; pp. 146) Harlow, U. K.

366. FLEISCHMANN, B. u. a. (Hrsg./Eds.; 1982): Operations Research Proceedings 1981. (Springer; 679 S., davon 119 in Englisch) Berlin, Heidelberg, New York

367. FLEISS, J. L. (1981): Statistical Methods for Rates and Proportions. 2nd ed. (Wiley; pp. 321) New York

368. FLIGNER, M. A. and G. E. POLICELLO II (1981): Robust rank procedures for the Behrens-Fisher problem. JASA **76**, 162-168

369. FLIGNER, M. A. and D. A. WOLFE (1982): Distribution-free tests for comparing several treatments with a control. STNE **36**, 119-127

370. FLURY, B. and H. RIEDWYL (1983): Angewandte Multivariate Statistik. Computergestützte Analyse mehrdimensionaler Daten. (G. Fischer; 187 S.) Stuttgart

371. FOERSTER, F. (1984): Computerprogramme zur Biosignalanalyse. (Springer; 141 S.) Berlin, Heidelberg, New York

372. FOLKS, J. L. and R. S. CHHIKARA (1978): The inverse Gaussian distribution and its statistical application - a review. With discussion. JRSB **40**, 263-289

373. FORD, B. L. and R. D. TORTORA (1978): A consulting aid to sample design. BICS **34**, 299-304

374. FORTHOFER, R. H. and R. G. LEHNEN (1981): Public Program Analysis. A New Approach to Categorical Data. (Lifetime Learning Publ. pp. 294) Belmont, Calif, 94002

375. FOSTER, H. D. (1980): Disaster Planning. The Preservation of Life and Property. (Springer; pp. 275) New York

376. FRANCIS, I. (1983): A survey of statistical software. COMP **1**, 17-27

377. FRANK, ILDICO E. and B. R. KOWALSKI (1982): Chemometrics. Analytical Chemistry **54**, 232R-243R

378. FRANKEL, E. G. (1984): Systems Reliability and Risk Analysis. (M. Nijhoff; pp. 423) The Hague

379. FREEMAN, D. H., Jr., Maria E. GONZALEZ, D. C. HOAGLIN and B. A. KILSS (1983): Presenting statistical papers. AMST **37**, 106-110

380. FREEMAN, D. H., Jr. and T. R. HOLFORD (1980): Summary rates. BICS **36**, 195-205

381. FREEMAN, P. R. (1983): The secretary problem and its extensions: a review. INTR **51**, 189-206

382. FREJKA, T. (1983): Weltbevölkerungsvorausschätzungen: Ein knapper geschichtlicher Überblick. Zeitschrift für Bevölkerungswissenschaft **9**, 73-92

383. FRENCH, S. and S. OATLEY (1982): Bayesian statistics: an overview. CRYS 19-51

384. FREY, J. H. (1983): Survey Research by Telephone. (Sage; pp. 201) Beverley Hills, Calif.

385. FROME, E. L. (1981): Poisson regression analysis. AMST **35**, 262-263

386. FROME, E. L. and H. CHECKOWAY (1985): Use of the Poisson regression models in estimating incidence rates and ratios. AMEP **121**, 309-323

387. FUJINO, Y. (1979): Tests for the homogeneity of a set of variances against ordered alternatives. BIKA **66**, 133-139

388. GAIL, M., R. WILLIAMS, D. P. BYAR and C. BROWN (1976): How many controls? CHRO **29**, 723-731

389. GAILE, G. L. and C. J. WILLMOTT (Eds.; 1984): Spatial Statistics and Models. (Reidel; pp. 460) Dordrecht and Boston

390. GALEN, R. S. und S. R. GAMBINO (1979): Norm und Normabweichung klinischer Daten. Der prädiktive Wert und die Effizienz von medizinischen Diagnosen. (G. Fischer; 195 S.) Stuttgart

391. GAMES, P. A. (1977): An improved t table for simultaneous control on g contrasts. JASA **72**, 531-534

392. GAMES, P. A., H. J. KESELMAN and Joanne C. ROGAN (1981): Simultaneous pairwise multiple comparison procedures for means when sample sizes are unequal. PYBU **90**, 594-598

393. GAMES, P. A., H. J. KESELMAN and J. C. ROGAN (1983): A review of simultaneous pairwise multiple comparisons. STNE **37**, 53-58

394. GANS, D. J. (1981): Corrected and extended tables for Tukey's quick test. TECS **23**, 193-195

395. GARDEN, J. S., D. G. Mitchell and W. N. Mills (1980): Nonconstant variance regression techniques for calibration-curve-based analysis. Analytical Chemistry **52**, 2310-2315

396. GART, J. J. (1978): The analysis of ratios and cross product ratios of Poisson variates with application to incidence rates. CSTH A **7**, 917-937

397. GART, J. J. and D. G. THOMAS (1982): The performance of three approximate confidence limit methods for the odds ratio. AMEP **115**, 453-470

398. GASS, S. I. (1983): Decision-aiding models: validation, assessment, and related issues for policy analysis. Operations Research **31**, 603-631

399. GAUS, W. (1983): Dokumentations- und Ordnungslehre. Lehrbuch für die Theorie und Praxis des Information Retrieval. (Springer; 504 S.) Berlin, Heidelberg, New York

400. GAVER, D. P. and Karen KAFADAR (1984): A retrievable recipe for inverse t. AMST **38**, 308-311

401. GAWRONSKI, W. and U. STADTMÜLLER (1981): Smoothing histograms by means of lattice and continuous distributions. MEKA **28**, 155-164

402. GEHRLEIN, W. V. and E. M. SANIGA (1978): Some exact critical values for the Kruskal-Wallis statistic. JQTE **10**, 73-75

[Geigy-Tabellen, Geigy Tables: see Nr./No. 1426]

403. GENTLE, J.E. (Ed.; 1984): Current Index to Statistics: Applications, Methods and Theory **9**, (1983), 1–577 IMSL, 7500 Bellaire, Houston, TX 77036 and Amer. Statist. Assoc., 806 15th Street, N.W., Washington, D.C. 20005

404. GENTLEMAN, Jane F. (Section Editor; 1984): New Developments in Statistical Computing. AMST **38**, 315–321 (e.g.)

405. GIBBONS, J.D. (1982): Methods for selecting the best process. JQTE **14**, 80–88

406. GIBBONS, J.D., I.OLKIN and M.SOBEL (1977): Selecting and Ordering Populations. A New Statistical Methodology. (Wiley; pp.569) New York (vgl. JASA **75**, (1980), 751–756)

407. GIBBONS, J.D., I.OLKIN and M.SOBEL (1979): An introduction to ranking and selection. AMST **33**, 185–195

408. GIBBONS, J.D. and J.W.PRATT (1975): P-values. Interpretation and methodology. AMST **29**, 20–25

409. GILCHRIST, R. (Ed.; 1982): GLIM 82. Proc. Internat. Conf. Generalized Linear Models, London 1982. (Lecture Notes in Statistics; Vol.14) (Springer; pp.188) Berlin, Heidelberg, New York and GILCHRIST, R., B.FRANCIS and J.WHITTAKER (Eds.; 1985): Generalized Linear Models. (Proc. of GLIM 85 Conf., Lancaster, UK, Sept.16–19, 1985) (Springer; pp.178) New York, Berlin, Heidelberg, Tokyo

410. GIRAULT, C. and W.REISIG (Eds.; 1982): Application and Theory of Petri Nets. (Informatik-Fachberichte, Bd.52) (Springer; pp.337) Berlin, Heidelberg, New York

411. GITTINS, R. (1985): Canonical Analysis. A Review with Applications in Ecology. (Springer; pp.320) Berlin, Heidelberg, New York, Tokyo

412. GLASS, G.V., P.D.PECKHAM and J.R.SANDERS (1972): Consequences of failure to meet assumptions underlying the fixed effects analysis of variance and covariance. Review of Educational Research **42**, 237–288

413. GLASS, G.V., V.L.WILLSON and J.M.GOTTMAN (1975): Design and Analysis of Time-Series Experiments. (Associated Univ. Press; pp.241) Boulder, Colorado

414. GLASS, G.V., B.McGAW and Mary L.SMITH (1981): Meta-Analysis in Social Research. (Sage; pp.278) Beverly Hills, Calif. and London

415. GLASTETTER, W., R.PAULERT und U.SPÖREL (1983): Die Entwicklung in der Bundesrepublik Deutschland 1950–1980. Befunde, Aspekte, Hintergründe. (Campus Vlg.; 614 S.) Frankfurt/New York

416. GLENN, N. (1977): Cohort Analysis. (Sage Publicat. Series 07-005; pp.72) Beverly Hills and London

417. GNANADESIKAN, R. (Ed.; 1983): Statistical Data Analysis. Proc. Symp. Appl. Math., Vol.28 (American Mathem. Soc.; pp.141) Providence, Rhode Island

418. GOLD, Ellen B. (1983): The Changing Risk of Disease in Women: An Epidemiologic Approach. (Collamore Press; pp.352) Lexington, MA 02173

419. GOLDBERG, J.D. and J.T.WITTES (1981): The evaluation of medical screening procedures. AMST **35**, 4-11

420. GOLDBERG, H. (1981): Extending the Limits of Reliability Theory. (Wiley; pp. 263) New York

421. GOLDER, E.R. and J.C.SETTLE (1976): The Box-Muller method for generating pseudo-random normal deviates. APST **25**, 12-20

422. GOLDSTEIN, H. (1979): The Design and Analysis of Longitudinal Studies. Their Role in the Measurement of Change. (Academic Press; pp. 199) London

423. GOLDSTEIN, M. and Inge F. GOLDSTEIN (1978): How We Know. An Exploration of the Scientific Process. (Plenum Press; pp. 357) New York

424. GOLLER, M. (Hrsg.; 1982): Simulationstechnik. (1.Symp., Erlangen, 26.-28. April 1982 Proc.) (Informatik-Fachber., Bd. 56) (Springer; 544 S.) Heidelberg, Berlin, New York

425. GOODMAN, L.A. (1978): Analyzing Qualitative/Categorical Data. Log-Linear Models and Latent-Structure Analysis. (Addison-Wesley; pp. 471) London, Amsterdam, Sydney, Tokyo

426. GOODMAN, L.A. (1979): Simple models for the analysis of association in cross-classifications having ordered categories. JASA **74**, 537-552

427. GOODMAN, L.A. (1981): Association models and the bivariate normal for contingency tables with ordered categories. BIKA **68**, 347-355

428. GOODMAN, L.A. (1983): The analysis of dependence in cross-classifications having ordered categories, using log-linear models for frequencies and log-linear models for odds. BICS **39**, 149-160

429. GOODMAN, L.A. and W.H.KRUSKAL (1979): Measures of Association for Cross Classifications. (Springer; pp. 146) New York

430. GOODNIGHT, J.H. and J.P.SALE (1982): SAS User's Guide: Statistics. (SAS Institute; pp. 584) Cary, N.C.

431. GOOVAERTS, M.J., F. DE VYLDER and J. HAEZENDONCK (1984): Insurance Premiums Theory and Applications. (North-Holland; pp. 406) Amsterdam

432. GOPAL, G.K. (Ed.; 1984): Papers presented at the Institute of Statisticians' Conference on Energy Statistics, Cambridge 1983. STAN **33**, 1-179

433. GORDON, A.D. (1979): A measure of the agreement between rankings. BIKA **66**, 7-15

434. GORDON, A.D. (1979): Another measure of the agreement between rankings. BIKA **66**, 327-332

435. GORDON, A.D. (1981): Classification. Methods for the Exploratory Analysis of Multivariate Data. (Chapman and Hall; pp. 193) London

436. GORE, A.P. and K.S.MADHAVA RAO (1982): Nonparametric test for slope in linear regression problems. BIJL **24**, 229-237

437. GORE, Sheila M., S.J.POCOCK and Gillian K.KERR (1984): Regression models and non-proportional hazards in the analysis of breast cancer survival. APST **33**, 176-195

438. Gower, J.C. (1982): The Yates algorithm. Utilitas Mathematica **21B**, 99-115

439. Graney, M.J. (1978): Nomogram for obtaining z-test statistic from Spearman r_s and sample size 10 to 1000. Teaching Sociology **5**, 201-210

440. Graney, M.J. (1979): Nomogram for obtaining z-test statistic from Kendall's S and sample size 10 to 50. Educational and Psychological Measurement **39**, 761-765

441. Gräsbeck, R., T. Alström and H.E. Solberg (Eds.; 1981): Reference Values in Laboratory Medicine. The Current State of the Art. (Based on a NORDKEM Workshop Hanaholmen, Espoo, Finland 4-6 May 1980) (Wiley; pp. 413) Chichester, New York

442. Gray, J.B. and R.F. Ling (1984): K-Clustering as a detection tool for influential subsets in regression. With discussion. TECS **26**, 305-330

443. Graybill, F. (1983): Matrices with Applications in Statistics. 2nd ed. (Wadsworth; pp. 480) Belmont, Calif.

444. Gregson, R.A.M. (1983): Time Series in Psychology. (L. Erlbaum; pp. 443) Hillsdale, N.J. and London

445. Green, P.E. (1976): Mathematical Tools for Applied Multivariate Analysis. (Academic Press; pp. 376) New York

446. Green, P.E. (1978): Analyzing Multivariate Data. (The Dryden Press; pp. 519) Hinsdale, Ill.

447. Green, R.H. (1979): Sampling Design and Statistical Methods for Environmental Biologists. (Wiley; pp. 257) New York

448. Greenacre, M. (1984): Theory and Applications of Correspondence Analysis. (Academic Press; pp. 376) New York and London

449. Greenland, S. (1985): An application of logistic models to the analysis of ordinal responses. BIJL **27**, 189-197

450. Greenland, S. and J.M. Robins (1985): Confounding and misclassification. AMEP **122**, 495-506

451. Griffiths, D. (1982): Letter to the Editor. STAN **31**, 268-270

452. Griffiths, D.A. (1980): Interval estimation for the three-parameter lognormal distribution via the likelihood function. APST **29**, 58-68

453. Griffiths, P. and I.D. Hill (Eds.; 1985): Applied Statistics Algorithms. (Wiley; pp. 307), Chichester and New York

454. Grimmet, G.R. and D.R. Stirzaker (1982): Probability and Random Processes. (Clarendon Press; pp. 354) Oxford

455. Gröner, G. (1982): Zur statistischen Erfassung der Wanderungsbewegung. Angew. Stat. Ökonom. **21**, 41-53

456. Grötschel, M., M. Jünger and G. Reinelt (1984): Optimal triangulation of large real world input-output matrices. Statistische Hefte **25**, 261-295

457. Gross, F. (1984): Placebo - the universal drug. MIME **23**, 176-182

458. Grossmann, W. (1985): Diskrimination und Klassifikation von Verlaufskurven. In: Pflug, G. Ch. (Hrsg.; 1985): Neuere Verfahren der nicht-

parametrischen Statistik. (Mediz. Informatik und Statist., Bd.60) (Springer; 129 S.) Berlin, Heidelberg, New York, Tokyo; S.109–127

459. GROVES, R. M. and R. L. KAHN (1979): Survey by Telephone – a national comparison with personal interviews. (Academic Press; pp.370) New York

460. GUENTHER, W. C. (1977): Sampling Inspection in Statistical Quality Control. (Griffin; pp.206) London

461. GUENTHER, W. C. (1977): Power and sample size determination when the alternative hypotheses are given in quantiles. AMST **31**, 117–118

462. GUENTHER, W. C. (1979): The use of noncentral F approximations for calculation of power and sample size. AMST **33**, 209–210

463. GUPTA, B.C. (1981): Checks on Yates' algorithm for a 2^m factorial experiment. Austral. J. Statist. **23**, 256–258

464. GUPTA, M. M. and E. SANCHEZ (Eds.; 1982): Fuzzy Information and Decision Processes. (North-Holland; pp.451) Amsterdam, New York, Oxford

465. GUPTA, S.S. and Woo-Chul KIM (1981): On the problem of selecting good populations. CSTH **10**, 1043–1077

466. GUPTA, S.S. and S. PANCHAPAKESAN (1979): Multiple Decision Procedures. Theory and Methodology of Selecting and Ranking Populations. (Wiley; pp.573) New York

467. GUPTA, S.S., S. PANCHAPAKESAN and J. K. SOHN (1985): On the distribution of the Studentized maximum of equally correlated normal random variables. CSSM **14**, 103–135

468. GUTTMAN, I. (1982): Linear Models. An Introduction. (Wiley-Interscience; pp.358) New York

469. GY, P. M. (1982): Sampling of Particulate Materials: Theory and Practice. (Elsevier; pp.431) New York

470. HABER, M. (1982): The continuity correction and statistical testing. INTR **50**, 135–144

471. HABER, M. (1984): Algorithm AS 207: Fitting a general loglinear model. APST **33**, 358–362

472. HABERMAN, S.J. (1978/79): Analysis of Qualitative Data: Vol.1, Introductory Topics; Vol.2, New Developments. (Academic Press; pp.612) New York and London

473. HACKING, I. (1975): The Emergence of Probability. A Philosophical Study of the Early Ideas about Probability, Induction and Statistical Inference. (Cambridge Univ. Pr.; pp.209) Cambridge

474. HÄRTLER, Gisela (1976): Versuchsplanung und statistische Datenanalyse. (Akademie-Verlag; 103 S.) Berlin

475. HÄRTLER, G. (1983): Statistische Methoden für die Zuverlässigkeitsanalyse. (Springer; 230 S.) Wien u. New York

476. HAHN, G.J. (1977): Fitting regression models with no intercept term. JQTE **9**, 56–61

477. HAHN, G.J. (1984): Experimental design in the complex world. TECS **26**, 19–31

478. HAHN, G.J. (1985): More intelligent statistical software and expert systems: future directions. With discussion. AMST **39**, 1–16

479. HAHN, G.J. and R.CHANDRA (1981): Tolerance intervals for Poisson and binomial variables. JQTE **13**, 100–110

480. HAHN, G.J. and W.Q.MEEKER, Jr. (1984): An engineer's guide to books on statistics and data analysis. JQTE **16**, 196–218

481. HAHN, G.J. and W.NELSON (1973): A survey of prediction intervals and their applications. JQTE **5**, 178–188

482. HAIR, J.F., Jr., R.E.ANDERSON, R.L.TATHAM and B.J.GRABLOWSKY (1984): Multivariate Data Analysis. With Readings. 2nd ed. (Petroleum Publ. Comp., PennWell; pp.360) Tulsa, Oklahoma 74101

483. HALD, A. (1981): Statistical Theory of Sampling Inspection by Attributes. (Academic Press; pp.518) London

484. HALDER, H.-R. und W.HEISE (1976): Einführung in die Kombinatorik. (Hanser; 302 S.) München und Wien

485. HALL, I.J. and D.D.SHELDON (1979): Improved bivariate normal tolerance regions with some applications. JQTE **11**, 13–19

486. HALL, P. (1982): Improving the normal approximation when constructing one-sided confidence intervals for binomial or Poisson parameters. BIKA **69**, 647–652

487. HAMER, G. (1981): Aufgaben und Probleme der amtlichen Statistik in der Bundesrepublik Deutschland aus ihren internationalen Kooperationsverpflichtungen. ASTA **65**, 40–61

488. HAMPEL, F. (1978): Modern trends in the theory of robustness. Math. Operationsforsch. und Stat., Ser. Statistics **9**, 425–442

489. HAN, C.-P. and J.L.G.WANG (1983): Computation of noncentral F distributions with even denominator degrees of freedom. CSSM (1983): **12**, 1–9

490. HAND, D.J. (1982): Discrimination and Classification. (Wiley; pp.224) New York

491. HARPER, W.M. and H.C.LIM (1982): Operational Research. 2nd ed. (Macdonald and Evans; pp.310) Plymouth

492. HARLOW, B.L., Jeanne F.ROSENTHAL and Regina G.ZIEGLER (1985): A comparison of computer-assisted and hard copy telephone interviewing. AMEP **122**, 335–340

493. HARRIS, Diana K. (1985): The Sociology of Aging. An Annotated Bibliography and Sourcebook. (Garland; pp.283) New York and London

494. HART, Anna E. (1983): The non-standard deviation. Teaching Statistics **5**, 16–20

495. HARTER, H.L. (1983): The Chronological Annotated Bibliography of Order Statistics. Vol.1: Pre-1950. Vol.2: 1950–59. (American Sciences Press; pp.515 and 993) Columbus, Ohio

496. HARTER, H.L. and D.B.OWEN (Eds.; 1973, 1974): Selected Tables in Mathematical Statistics, 1 and 2. (Institute of Mathematical Statistics, Am. Math. Soc.; pp. 403; 388) Providence, Rhode Island

497. HARTGE, Patricia, Louise A.BRINTON, Jeanne F.ROSENTHAL, J.I.CA-HILL, R.N.HOOVER and J.WAKSBERG (1984): Random digit dialing in selecting a population-based control group. AMEP **120**, 825–833

498. HARTIGAN, J.A. (1983): Bayes Theory. (Springer; pp. 145) New York, Heidelberg, Berlin

499. HARTLEY, H.O. (1980): Statistics as a science and as a profession. JASA **75**, 1–7

500. HARTMANN, D.P., J.M.GOTTMAN, R.R.JONES, W.GARDNER, A.E.KAZ-DIN and R.S.VAUGHT (1980): Interrupted time-series analysis and its application to behavioral data. Journal of Applied Behavior Analysis **13**, 543–559

501. HARTUNG, J. und Bärbel ELPELT (1984): Multivariate Statistik. Lehr- und Handbuch der angewandten Statistik. (R.Oldenbourg; 806 S.) München und Wien

502. HARTWIG, F. and B.E.DEARING (1979): Exploratory Data Analysis. (Sage Publicat. Series 07-016; pp. 83) Beverley Hills and London

503. HASSELBLAD, V., A.G.STEAD and W.GALKE (1980): Analysis of coarsely grouped data from the lognormal distribution. JASA **75**, 771–778

504. HASTINGS, N.A.J. and J.B.PEACOCK (1979): Statistical Distribution. (Wiley; pp. 130) New York

505. HAUSER, J.A. (1982): Bevölkerungslehre für Politik, Wirtschaft und Verwaltung. (UTB Nr. 1164) (P.Haupt; 359 S.) Bern und Stuttgart

506. HAUSS, K. (Hrsg.; 1981): Medizinische Psychologie im Grundriß. 2. erw. Aufl. (Vlg. f. Psychologie, Dr. C.J.Hogrefe; 532 S.) Göttingen, Toronto, Zürich (insbes. S. 203/234)

507. HAUX, R. (1983/84): Statistical analysis systems. Construction and aspects of method design. SSNL **9**, 106–115 and **10**, 14–27

508. HAUX, R. (1985): Datenbankaspekte bei statistischen Auswertungssystemen. EDVM **16**, 41–46

509. HAUX, R., M.SCHUMACHER and G.WECKESSER (1984): Rank tests for complete block designs. BIJL **26**, 567–582

510. HAWKES, A.G. (1982): Approximating the normal tail. STAN **31**, 231–236

511. HAWKINS, D.M. (1980): Identification of Outliers. (Chapman and Hall; pp. 224) London

512. HAWKINS, D.M. (1981): Testing for equality of variances of correlated normal variables. STNE **35**, 39–47

513. HAWKINS, D.M. (Ed.; 1982): Topics in Applied Multivariate Analysis. (Cambridge Univ. Press; pp. 362) Cambridge

514. HAYNAM, G.E., Z. GOVINDARAJULU, F.C. LEONE and P. SIEFERT (1982/83): Tables of the cumulative non-central chi-square distribution

- parts 1-4. Math. Operationsforsch. Statist., Ser. Statistics **13**, 413-443, 577-634, **14**, 75-139, 269-300

515. HEALY, M.J.R. (1979): Outliers in clinical chemistry quality control schemes. Clinical Chemistry **25**, 675-677

516. HEARNE, E.M., III and G.M.CLARK (1983): Validity and power investigation of a test for serial correlation in univariate repeated measures analyses. SMED **2**, 497-504

517. HEIBERGER, R.M. (1981): The specification of experimental designs to ANOVA programs. With comments and reply. AMST **35**, 98-108

518. HEILER, S. (1982): Zeitreihenanalyse heute: Ein Überblick. ASTA **65**, 376-402

519. HEILIG, G. (1983): Bildung und Anwendung nichtmetrischer Modelle in der Bevölkerungswissenschaft (GSK-Ansatz). Zeitschrift für Bevölkerungswissenschaft **9**, 447-474

520. HEINER, K.W., J.W. WILKINSON and R.S. SACHER (Eds.; 1983): Computer Science and Statistics: Proceedings of the 14th Symposium on the Interface. (Springer; pp.313) New York, Heidelberg, Berlin

521. HEINSOHN, G., R.KNIEPER und O.STEIGER (1979): Menschenproduktion. Allgemeine Bevölkerungstheorie der Neuzeit. (Suhrkamp [ed. s.914]; 258 S.) Frankfurt/M.

522. HEISE, D.R. (1975): Causal Analysis. (Wiley-Interscience; pp.301) New York

523. HENDRY, D.F. and J.-F.RICHARD (1983): The econometric analysis of economic time series. With discussion. INTR **51**, 111-163

524. HENLEY, E.J. and H.KUMAMOTO (1980): Reliability Engineering and Risk Assessment. (Prentice-Hall; pp.568) New York

525. HENNING, H.J. und K.MUTHIG (1979): Grundlagen konstruktiver Versuchsplanung. Ein Lehrbuch für Psychologen. (Kösel-Vlg.; 263 S.) München

526. HENSCHEL, H. (1979): Wirtschaftsprognosen. (F.Vahlen; 156 S.) München

527. HERSEN, M. and D.H.BARLOW (and A.E.KAZDIN; 1976): Single-Case Experimental Designs: Strategies for Studying Behavior Change. (Pergamon Press; pp.374) New York, ..., Frankfurt

528. HERTZ, D. and H.THOMAS (1983): (1) Risk Analysis and Its Applications. (2) Practical Risk Analysis, an Approach through Case Histories. (Wiley; pp.246, 360) New York

529. HERZOG, O., W.REISIG und R.VALK (1984): Petri-Netze: ein Abriß ihrer Grundlagen und Anwendungen. Informatik-Spektrum **7**, 20-27

530. HEWETT, J.E. and Z. LABABIDI (1982): Comparison of three regression lines over a finite interval. BICS **38**, 837-891

531. HEWLETT, P.S. and R.L PLACKETT (1979): An Introduction to the Interpretation of Quantal Responses in Biology. (E.Arnold; pp. 82) London

532. HEYDE, C.C. (1982): Trends in the statistical sciences. Australian Journal of Statistics **23**, 273-286

533. HILGERS, R. (1982): On the Wilcoxon-Mann-Whitney-test as nonparametric analogue and extension of t-test. BIJL **24**, 3-15

534. HILL, D.L. and P.V. RAO (1981): Tests of symmetry based on Watson statistics. CSTH **10**, 1111-1125

535. HILL, M.O. (1974): Correspondence analysis: a neglected multivariate method. APST **23**, 340-354

536. HILLIER, F.S. and G.J. LIEBERMAN (1980): Introduction to Operations Research. 3rd ed. (Holden-Day; pp. 829) Oakland, Calif.

537. HILLS, M. (1967): Discrimination and allocation with discrete data. APST **16**, 237-250

538. HINKELMANN, K. (Ed.; 1984): Experimental Design, Statistical Models and Genetic Statistics. (M. Dekker; pp. 409) New York

539. HINSCHÜTZER, Ursula und Heide MOMBER (1982): Basisdaten über ältere Menschen in der Statistik der Bundesrepublik Deutschland. (Deutsches Zentrum für Altersfragen e. V., 457 S.) 1000 Berlin 42

540. HINZ, P. and J. GURLAND (1967): Simplified techniques for estimating parameters of some generalized Poisson distributions. BIKA **54**, 555-566

541. HIRANO, K., H. KUBOKI, S. AKI and A. KURIBAYASHI (1983): Figures of Probability Density Functions in Statistics I. Continuous Univariate Case. (Computer Science Monograph No. 19) (The Institute of Statistical Mathematics; pp. 87) Tokyo

542. HIRANO, K., H. KUBOKI, S. AKI and A. KURIBAYASHI (1984): Figures of Probability Density Functions in Statistics II. Discrete Univariate Case. (Computer Science Monograph No. 20) (The Institute of Statistical Mathematics; pp. 199) Tokyo

543. HITCHING, F. (1982): Die letzten Rätsel unserer Welt. Übers. aus dem Engl. von Pro Interpret GmbH. (Umschau-Vlg.; 296 S.) Frankfurt am Main

544. HJORTH, U. (1980): A reliability distribution with increasing, decreasing, constant and bathtub-shaped failure rates. TECS **22**, 99-107

545. HOAGLIN, D.C. (1977): Direct approximations for chi-squared percentage points. JASA **72**, 508-515

546. HOAGLIN, D.C. (1980): A Poissonness plot. AMST **34**, 146-149

547. HOAGLIN, D.C., F. MOSTELLER and J.W. TUKEY (Eds.; 1983): Understanding Robust and Exploratory Data Analysis. (Wiley; pp. 447) New York

548. HOAGLIN, D.C. and R. WELSCH (1978): The hat matrix in regression and ANOVA. AMST **32**, 17-22

549. HOCKING, R.R. (1976): The analysis and selection of variables in linear regression. BICS **32**, 1-49

550. HOCKING, R.R. (1983): Developments in linear regression methodology: 1959-1982. With discussion. TECS **25**, 219-249

551. HOCKING, R.R. and O.J. PENDLETON (1983): The regression dilemma. CSTH **12**, 497-527

552. HÖFLE-ISPHORDING, U. (1978): Zuverlässigkeitsrechnung. Einführung in ihre Methoden. (Springer; 179 S.) Heidelberg

553. HÖHN, Charlotte, U. MAMMEY und K. SCHWARZ (1981): Die demographische Lage in der Bundesrepublik Deutschland. Zeitschr. f. Bevölkerungswissenschaft 7, 139-230

554. HÖLDER, E. (1984): Das Statistische Bundesamt. Das Wirtschaftsstudium 13, 401-405 und 431

555. HOERL, A. E. and R. W. KENNARD (1970): Ridge regression: biased estimation for nonorthogonal problems. TECS 12, 55-67

556. HOERL, R. W. (1985): Ridge analysis 25 years later. AMST 39, 186-192

557. HOFF, J. C. (1983): A Practical Guide to Box-Jenkins Forecasting. (Lifetime Learning Publications; pp. 316) Belmont, Calif. 94002

558. HOGARTH, R. M. (Ed.; 1982): Question Framing and Response Consistency. (Jossey-Bass; pp. 109) San Francisco, Calif.

559. HOGG, R. V. and R. V. LENTH (1984): A review of some adaptive statistical techniques. CSTH 13, 1551-1579

560. HOGG, Sheilah A. and A. CIAMPI (1985): GFREG: A computer program for maximum likelihood regression using the generalized F distribution. Computer Methods and Programs in Biomedicine 20, 201-215

561. HOGUE, Carol J. R., D. W. GAYLOR and K. F. SCHULZ (1983): Estimators of relative risk for case-control studies. AMEP 118, 396-407

562. HOLFORD, T. R. (1980): The analysis of rates and of survivorship using log-linear models. BICS 36, 299-305

563. HOLLER, M. J. (Hrsg.; 1984): Wahlanalyse. Hypothesen, Methoden und Ergebnisse. (tuduv Vlg.; 230 S.) München 2, Gabelsbergerstr. 15

564. HOLM, K. (Hrsg.; 1979): Die Befragung 6. (UTB Nr. 436) (A. Francke; 277 S.) München

565. HOLM, S. (1979): A simple sequentially rejective multiple test procedure. Scandinavian Journal of Statistics 6, 65-70

566. HOLMES, D. I. (1983): A graphical identification procedure for growth curves. STAN 32, 405-415

567. HOMMEL, G. (1985): Multiple Vergleiche mittels Rangtests - alle Paarvergleiche. In: Pflug, G. Ch. (Hrsg.; 1985): Neuere Verfahren der nichtparametrischen Statistik. (Mediz. Informatik und Statist., Bd. 60) (Springer; 129 S.) Berlin, Heidelberg, New York, Tokyo, S. 28-48

568. HOOKE, R. (1983): How to Tell the Liars from the Statisticians. (M. Dekker; pp. 173) New York

569. HORA, S. C. and G. D. KELLEY (1983): Bayesian inference on the odds and risk ratios. CSTH 12, 725-738

570. HORN, M. (1981): On the theory for RYANs universal multiple comparison producedure with treatment of ties in the ranks of samples. BIJL 23, 343-355

571. HORNE, G. P. and M. C. K. YANG (1982): Time series analysis for single-subject designs. PYBU 91, 178-189

572. Hornung, J. (1977): Kritik des Signifikanztests. Metamed **1**, 325–345

573. Horton, R. L. (1978): The General Linear Model. Data Analysis in the Social and Behavioral Sciences. (McGraw-Hill; pp. 274) New York

574. Hosmer, D. W. and S. C. Hartz (1981): Methods for analyzing odds ratios in a $2 \times c$ contingency table. BIJL **23**, 741–748

575. Hossack, I. B., J. H. Pollard and B. Zehnwirth (1983): Introductory Statistics with Applications in General Insurance. (Cambridge Univ. Press; pp. 280) Cambridge

576. Howe, D. R. (1983): Data Analysis for Data Base Design. (E. Arnold; pp. 307) London

577. Hubaux, A. and G. Vox (1970): Decision and detection limits for linear calibration curves. Analytical Chemistry **42**, 849–855

578. Huber, H. P. (1980): Zur Auswertung mehrfaktorieller Rangvarianzanalysen bei ungleichen Zellbesetzungen. Teil I: Versuchspläne mit unabhängigen Stichproben. PYBE **22**, 553–573

579. Huber, H. P. (1982): Zur Auswertung mehrfaktorieller Rangvarianzanalysen bei ungleichen Zellbesetzungen. Teil II: Versuchspläne mit abhängigen Stichproben. PYBE **24**, 419–446

580. Huber, P. J. (1981): Robust Statistics. (Wiley; pp. 308) New York

581. Huber, R. K. (Ed.; 1984): Systems Analysis and Modeling in Defense. Development, Trends, Issues. (Plenum Publ. Corp.; pp. 930) New York

582. Hubert, J. J. (1980): Bioassay. (Kendall/Hunt Publ. Co.; pp. 164) Dubuque, Iowa

583. Huberty, C. J. (1984): Issues in the use and interpretation of discriminant analysis. PYBU **95**, 156–171

584. Hübner, E. und H.-H. Rohlfs (1984): Jahrbuch der Bundesrepublik 1984. (Beck/dtv [TB Nr. 5250]; 521 S.) München

585. Huitema, B. E. (1980): The Analysis of Covariance and Alternatives. (Wiley; pp. 440) New York

586. Huitson, A., J. Poloniecki, R. Hews and N. Barker (1982): A review of cross-over trials. STAN **31**, 71–80 [see also P. Armitage and M. Hills: 119–131]

587. Humak, K. M. S. (1977/84): Statistische Methoden der Modellbildung I–III. I.: Statistische Inferenz für lineare Parameter. II.: Nichtlineare Regression, robuste Verfahren in linearen Modellen, Modelle mit Fehlern in den Variablen. III.: Statistische Inferenz für Kovarianzparameter. (Akademie-Verlag; 516, 360, 344 S.) Berlin

588. Hunter, J. E., F. L. Schmidt and G. B. Jackson (1982): Meta-Analysis. Cumulating Research Findings Across Studies. (Sage; pp. 176) Beverly Hills, Calif. and London

589. Hunter, W. G. and W. F. Lamboy (1981): A Bayesian analysis of the linear calibration problem. With discussion. TECS **23**, 323–350

590. Hussain, S. S. and P. Sprent (1983): Non-parametric regression. JRSA **146**, 182–191

591. HUSSMANS, R., U. MAMMEY und R. SCHULZ (1983): Die demographische Lage in der Bundesrepublik Deutschland. Zeitschrift für Bevölkerungswissenschaft **3**, 291–362

592. HWANG, F. K. (1984): Robust group testing. JQTE **16**, 189–195

593. IACHAN, R. (1982): Systematic sampling: a critical review. INTR **50**, 293–303

594. IACHAN, R. (1983): Nonsampling errors in surveys – A review. CSTH **12**, 2273–2287

595. IACHAN, R. (1984): Sampling strategies, robustness and efficiency: the state of the art. INTR **52**, 209–218

596. IHM, P. (1984): Korrespondenzanalyse und Gaußsches Ordinationsmodell. ASTA **68**, 41–62

597. IMAN, R. L. and W. J. CONOVER (1978): Approximations of the critical region for Spearman's rho with and without ties present. CSSM B 7, 269–282

598. IMAN, R. L. and J. M. DAVENPORT (1976): New Approximations to the exact distribution of the Kruskal-Wallis test statistic. CSTH A **5**, 1335–1348

599. IMAN, R. L. and J. M. DAVENPORT (1980): Approximations of the critical region of the Friedman statistic. CSTH A **9**, 571–595

600. IMAN, R. L., D. QUADE and D. A. ALEXANDER (1975): Exact probability levels for the Kruskal-Wallis test. Selected Tables in Mathematical Statistics **3**, 329–384

601. IMPAGLIAZZO, J. (1985): Deterministic Aspects in Mathematical Demography. (Biomathematics, Vol. 13) (Springer; pp. 200) New York, Heidelberg

602. INSKIP, Hazel, Valerie BERAL and Patricia FRASER (1983): Methods for age-adjustment of rates. SMED **2**, 455–466

603. JACKSON, D. J. and E. F. BORGATTA (Eds.; 1981): Factor Analysis and Measurement in Sociological Research. A Multi-Dimensional Perspective. (Sage Publicat. Series 07-021; pp. 313) Beverly Hills, Calif. 90212

604. JACKSON, J. E. (1980/81): Principal components and factor analysis. JQTE **12**, 201–213; **13**, 46–58 and 125–130

605. JAECH, J. L. (1979): Estimating within-laboratory variability from interlaboratory test data. JQTE **11**, 185–191

606. JAFFE, A. (1984): Ordering the universe: the role of mathematics. SIAM Review **26**, 473–500

607. JAIN, R. B. (1981): Detecting outliers: power and some other considerations. CSTH A **10**, 2299–2314

608. JAMBU, M. and M.-O. LEBEAUX (1983): Cluster Analysis and Data Analysis. (North-Holland; pp. 898) Amsterdam

609. JANDOK, W. (1982): Methodische Grundfragen eines rationellen Entscheidungsablaufs bei Fragebogenanwendungen in der Medizin. ZGHY **28**, 878–881 [vgl. auch **30**, (1984), 242–245]

610. JANSEN, A. A. M. (1980): Comparative calibration and congeneric measurements. BICS **36**, 729-734

611. JANSEN, Marlies E. (1984): Ridit analysis, a review. STNE **38**, 141-158

612. JANSSEN, J., J.-F. MARCOTORCHINO and J.-M. PROTH (Eds.; 1983): New Trends in Data Analyses and Applications. (North-Holland; pp. 274) Amsterdam

613. JASSO, Guillermina (1980): A new theory of distributive justice. American Sociological Review **45**, 3-32

614. JAZAIRI, N. T. (1983): The present state of the theory and practice of index numbers. Bull. Int. Statist. Inst. **50**, 122-147

615. JEFFERS, J. N. R. (1982): Modelling. Outline Studies in Ecology. (Chapman and Hall; pp. 80) London and New York

616. JEFFERS, J. N. R. (1982): Forestry biometry – a review. Statist. Theory Practice (Umeå, Sweden), 305-325

617. JEGER, M. (1973): Einführung in die Kombinatorik. Band 1. (Klett; 218 S.) Stuttgart

618. JELLINEK, G. (1981): Sensorische Lebensmittelprüfung. Lehrbuch für die Praxis. (D&PS-Vlg.; 585 S.) Pattensen bei Hannover

619. JENSEN, E. B., A. J. BADDELEY, H. J. G. GUNDERSEN and R. SUNDBERG (1985): Recent trends in sterology. INTR **53**, 99-108

620. JESDINSKY, H. J. und J. TRAMPISCH (Hrsg.; 1985): Prognose- und Entscheidungsfindung in der Medizin. (Mediz. Informatik und Statist., Bd. 62) (Springer; pp. 524) Berlin, Heidelberg, New York, Tokyo

621. JESSEN, R. J. (1978): Statistical Survey Techniques. (Wiley; pp. 520) New York

622. JILEK, M. (1981): A bibliography of statistical tolerance regions. Mathematische Operationsforschung und Statistik, ser. statistics **12**, 441-456

623. JILEK, M. (1982): Sample size and tolerance limits. Trab. Estadist. **33**, 64-78

624. JILEK, M. (1982): On errors of measurement. BIJL **24**, 493-501

625. JOE, H., J. A. KOZIOL and A. J. PETKAU (1981): Comparison of procedures for testing the equality of survival distributions. BICS **37**, 327-340

626. JÖCKEL, K.-H. und P. PFLAUMER (1981): Demographische Anwendungen neuerer Zeitreihenverfahren. Zeitschrift für Bevölkerungswissenschaft **7**, 519-542

627. JÖCKEL, K.-H. und B. WOLTER (1984): Statistische Überlegungen zur Verlaufskurvenanalyse. EDVM **15**, 100-106

628. JÖRESKOG, K. G. (1981): Analysis of covariance structures. With discussion. Scandinavian Journal of Statistics **8**, 65-92

629. JOHN, R. D. and J. ROBINSON (1983): Significance levels and confidence intervals for permutation tests. JSCS **16**, 161-173

630. JOHNSON, M. E. and V. W. LOWE (1979): Bounds on the sample skewness and kurtosis. TECS **21**, 377-378

631. JOHNSON, N. L. and S. KOTZ (1969, 1970, 1972): Distributions in Statistics. I. Discrete Distributions. II. Continuous univariate Distributions, 1 and 2. III. Continuous Multivariate Distributions. (Houghton Mifflin and Wiley; pp. 328; 300, 306; 333) New York

632. JOHNSON, N. L. and S. KOTZ (1982): Developments in discrete distributions, 1969-1980. INTR **50**, 71-101

633. JOHNSON, N. L. and B. L. WELCH (1940): Applications of the noncentral t-distribution. BIKA **31**, 362-389

634. JOHNSON, R. A. and D. W. WICHERN (1982): Applied Multivariate Statistical Analysis. (Prentice-Hall; pp. 594) Englewood Cliffs, N. J.

635. JOINER, B. L. (1981): Lurking variables: some examples. AMST **35**, 227-233

636. JONCKHEERE, A. R. (1954): A distribution-free k-sample test against ordered alternatives. BIKA **41**, 133-145

637. JONES, A. J. (1979): Game Theory. Mathematical Models of Conflict. (Halstead Press, Wiley; pp. 309) New York

638. JONES-LEE, M. W. (Ed.; 1982): The Value of Life and Safety. (North-Holland; pp. 310) Amsterdam

639. KÄHLER, W.-M. (1984): Einführung in das Datenanalysesystem SPSS. (Vieweg; 215 S.) Braunschweig, Wiesbaden

640. KAHNEMANN, D., P. SLOVIC and A. TVERSKY (Eds.; 1982): Judgment under Uncertainty: Heuristics and Biases. (Cambridge Univ. Press; pp. 555) Cambridge

641. KAILA, K. L. (1980): A new method for linear curve fitting with both variables in error. SANB **42**, 81-97

642. KALBFLEISCH, J. D. and R. L. PRENTICE (1980): The Statistical Analysis of Failure Time Data. (Wiley; pp. 321) New York

643. KALBFLEISCH, J. G. (1979): Probability and Statistical Inference. I, II. (Springer; pp. 342, 316) New York [2nd ed., pp. 343, 360, 1985]

644. KALTON, G. (1983): Models in the practice of survey sampling. INTR **51**, 175-188

645. KALTON, G. (1983): Introduction to Survey Sampling. (Sage Publicat. Series 07-035; pp. 96) Beverly Hills and London

646. KALTON, G. and H. SCHUMAN (1982): The effect of the question on survey responses: a review. With discussion. JRSA **145**, 42-73

647. KANE, V. E. (1982): Standard and goodness-of-fit parameter estimation methods for the three-parameter lognormal distribution. CSTH **11**, 1935-1957

648. KANER, H. C., S. G. MOHANTY and J. C. LYONS (1980): Critical values of the Kolmogorov-Smirnov one-sample tests. PYBU **88**, 498-501

649. KARDAUN, O. (1983): Statistical survival analysis of male larynx-cancer patients - a case study. STNE **37**, 103-125

650. KARLIN, S., T. AMEMIYA and L. A. GOODMAN (Eds.; 1983): Studies in Econometrics, Time Series and Multivariate Statistics. (Academic Press; pp. 595) New York and London

651. KAY, R. (1977): Proportional hazard regression models and the analysis of censored survival data. APST **26**, 227-237

652. KEATING, J. P. and O. L. HENSLEY (1983): A note on the critical values of the *F*max-test. CSSM **12**, 257-263

653. KEEL, A. (1982): Eine Verallgemeinerung des exakten Fisherschen Unabhängigkeitstests von 2 auf *r* dichotome Merkmale. Statistische Hefte **23**, 333-340

654. KEENEY, R. L. (1982): Decision analysis: an overview: Operations Research **30**, 803-838

655. KEMPTHORNE, O. (1976): Of what use are tests of significance and tests of hypothesis. CSTH A **5**, 763-777

656. KENDALL, M. G. (1968): Introduction to model building and its problems. First paper in A. M. W. (?; Ed.): Mathematical Building in Economics and Industry. (CEIR Ltd., Conf. London July 1967) (Griffin; pp. 165) London, pp. 1-14

657. KENDALL, M. G. and W. R. BUCKLAND (1982): A Dictionary of Statistical Terms. 4th ed., revised and enlarged (Longman; pp. 213) London and New York

658. KENDALL, M. G., A. STUART and J. K. ORTH (1984): The Advanced Theory of Statistics: Vol. 1. Distribution Theory. Vol. 2. Inference and Relationship. Vol. 3. Design and Analysis and Time Series. Fourth edition. (Griffin; pp. 484, 758, 790) London

659. KENETT, R. S. (1983): On an exploratory analysis of contingency tables. STAN **32**, 395-403

660. KENNEDY, J. J. (1983): Analyzing Qualitative Data. (Praeger; pp. 284) New York

661. KENNEDY, W. J., Jr. and J. E. GENTLE (1980): Statistical Computing. (M. Dekker; pp. 591) New York

662. KENNY, D. A. (1979): Correlation and Causality. (Wiley; pp. 288) New York

663. KEREN, G. (Ed.; 1982): Statistical and Methodological Issues in Psychology and Social Sciences Research. (Lawrence Erlbaum Associates; pp. 390) Hillsdale, New Jersey

664. KESELMAN, H. J. and Joanne C. ROGAN (1977): The Tukey multiple comparison test: 1953-1976. PYBU **84**, 1050-1056

665. KESSLER, R. C. and D. F. GREENBERG (1981): Linear Panel Analysis. Models of Quantitative Change. (Academic Press; pp. 203) New York and London

666. KEULS, M. and F. GERRETSEN (1982): Statistical analysis of growth curves in plant breeding. Euphytica **31**, 51-64

667. KEYFITZ, N. and J. A. BEEKMAN (1983): Demography through Problems. (Springer; pp. 240) New York, Heidelberg, Berlin

668. KEYFITZ, N. and G. LITTMAN (1979): Mortality in a heterogeneous population. Population Studies **33**, 333-342

669. KHAMIS, S. H. (1983): Applications of index numbers in international comparisons and related concepts. Bull. Int. Statist. Inst. **50**, 171-188

670. KIM, W., D. S. REINER and D. S. BATORY (Eds.; 1984): Query Processing in Database Systems. (Springer; pp. 339) Berlin, Heidelberg, New York

671. KINCANNON, C. L. (Ed.; 1983): Statistical Abstract of the United States 1984. (US Dept. of Commerce, Bureau of the Census; pp. 1042) Washington, D. C.

672. KING, M. L. (1981): The Durbin-Watson test for serial correlation: bounds for regressions with trend and/or seasonal dummy variables. Econometrica **49**, 1571-1581

673. KING, M. L. (1983): The Durbin-Watson test for serial correlation. Bounds for regressions using monthly data. J. Econometrics **21**, 357-366

674. KIRK, R. E. (1982): Experimental Design: Procedures for the Behavioral Sciences. 2nd ed. (Wadsworth; pp. 911) Belmont, Calif.

675. KISH, L. (1978): Chance, statistics and statisticians. JASA **73**, 1-6

676. KISH, L. and V. VERMA (1983): Census plus samples: Combined uses and designs. Bull. Int. Statist. Inst. **50**, 66-82

677. KITAGAWA, Evelyn M. (1955): Components of difference between two rates. JASA **50**, 1168-1194

678. KITAGAWA, Evelyn M. (1964): Standardized comparisons in population research. Demography **1**, 296-315

679. KLÄY, M. und H. RIEDWYL (1984): ALSTAT 1: Algorithmen der Statistik für Kleinrechner. ALSTAT 2: Algorithmen der Statistik für Hewlett-Packard HP-41C. (Birkhäuser; 248 und 171 S.) Basel, Boston, Stuttgart

680. KLECKA, W. R. (1980): Discriminant Analysis. (Sage Publicat. Series 07-019; pp. 70) Beverly Hills and London

681. KLEINBAUM, D. G. and L. L. KUPPER (1978): Applied Regression Analysis and Other Multivariable Methods. (Duxbury Press; pp. 556) North Scituate, Mass.

682. KLEINBAUM, D. G., L. L. KUPPER and H. MORGENSTERN (1982): Epidemiologic Research: Principles and Quantitative Methods. (Lifetime Learning Publications; pp. 529) Belmont, Calif.

683. KLEINER, B. and J. A. HARTIGAN (1981): Representing points in many dimensions by trees and castles. JASA **76**, 260-269 [Comments and rejoinder 269-295]

684. KLEITER, G. B. (1981): Bayes-Statistik. Grundlagen und Anwendungen. (de Gruyter; 569 S.) Berlin

685. KNAFL, G., C. SPIEGELMAN, J. SACKS and D. YLVISAKER (1984): Nonparametric calibration. TECS **26**, 233-241

686. KNAPP, T. R. (1978): Canonical correlation analysis. A general parametric significance-testing system. PYBU **85**, 410-416

687. KNAPP, T. R. (1982): The birthday problem: some empirical data and some approximations. Teaching Statistics **4**, 10-14

688. KNOKE, J. D. (1977): Testing for randomness against autocorrelation: alternative tests. BIKA **64**, 523-529

689. Knoke, J. D. (1982): Discriminant analysis with discrete and continuous variables. BICS **38**, 191–200

690. Koch, G. G., I. A. Amara, M. E. Stokes and D. B. Gillings (1980): Some views on parametric and non-parametric analysis for repeated measurements and selected bibliography. INTR **48**, 249–265

691. Köchel, P. (1983): Zuverlässigkeit technischer Systeme. (H. Deutsch; 156 S.) Thun und Frankfurt/Main [bzw. VEB Fachbuchverlag Leipzig]

692. Köhler, C. O., P. Tautu und G. Wagner (Hrsg.; 1984): Der Beitrag der Informationsverarbeitung zum Fortschritt der Medizin. (Mediz. Informatik und Statistik, Bd. 50) (Springer; 668 S.) Berlin, Heidelberg, New York, Tokyo

693. Koehler, K. J. (1983): A simple approximation for the percentiles of the t distribution. TECS **25**, 103–105

694. Köpcke, W. (1984): Zwischenauswertungen und vorzeitiger Abbruch von Therapiestudien. (Mediz. Informatik und Statist., Bd. 53) (Springer; 197 S.) Berlin, Heidelberg, New York, Tokyo

695. Kohlas, J. (1977): Stochastische Methoden des Operations Research. (Teubner; 192 S.) Stuttgart

696. Konheim, A. G. (1981): Cryptography. A Primer. (Wiley; pp. 450) New York

697. Koopman, P. A. R. (1984): Confidence intervals for the ratio of two binomial proportions. BICS **40**, 513–517

698. Korhonen, M. P. (1982): On the performance of some multiple comparison procedures with unequal variances. Scand. J. Statist. **9**, 241–247

699. Koslow, B. A. und I. A. Uschakow (1979): Handbuch zur Berechnung der Zuverlässigkeit für Ingenieure. (Hanser; 598 S.) München

700. Kotz, S. and N. L. Johnson (1984): Effects of false and incomplete identification of defective items on the reliability of acceptance sampling. Operations Research **32**, 575–583

701. Kotz, S., N. L. Johnson and C. B. Read (Eds.; 1982–1986): Encyclopedia of Statistical Sciences. Vol. I–VIII. (Wiley; approximately 5000 pages) New York

702. Kowalski, C. J. and K. E. Guire (1974): Longitudinal data analysis (with a selected bibliography of 225 entries and 11 classifications). Growth **38**, 131–169

703. Koziol, J. A. (1984): VARCOV: A computer program for the distribution-free analysis of growth and response curves. Computer Programs in Biomedicine **19**, 69–74

704. Koziol, J. A. and M. D. Perlman (1978): Combining independent chi-squared tests. JASA **73**, 753–763

705. Kracke, H. (1983): Mathe-musische Knobelisken. Tüfteleien für Tüftler und Laien. 2. durchges. Aufl. (Dümmler; 445 S.) Bonn

706. Kraemer, Helena C. (1980): Extension of the kappa coefficient. BICS **36**, 207–216

707. KRÄMER, H. L. (1983): Soziale Schichtung. Einführung in die moderne Theoriediskussion. (M. Diesterweg; 140 S.) Frankfurt/M., Berlin, München

708. KRAUTH, J. (1980): Nonparametric analysis of response curves. Journal of Neuroscience Methods **2**, 239-252

709. KRAUTH, J. (1982): Verteilungsfreie Homogenitätstests bei abhängigen Stichproben. PYBE **24**, 601-619 [vgl. auch **26** (1984) 309-317]

710. KRAUTH, J. (1985): A comparison of tests for marginal homogeneity in square contingency tables. BIJL **27**, 3-15

711. KREMER, E. (1985): Einführung in die Versicherungsmathematik. (Vandenhoeck und Ruprecht; 159 S.) Göttingen

712. KRES, H. (1983): Statistical Tables for Multivariate Analysis. A Handbook with References to Applications. (Springer; pp. 504) New York, Berlin, Heidelberg, Tokyo

713. KREWSKI, D. and C. BROWN (1981): Carcinogenic risk assessment. A guide to the literature. BICS **37**, 353-366

714. KREWSKI, D., R. PLATEK and J. N. K. RAO (Eds.; 1981): Current Topics in Survey Sampling. (Academic Press; pp. 509) New York

715. KRIPPENDORFF, K. (1980): Clustering. Chapter 9 in P. R. Monge and J. N. Cappella (Eds.): Multivariate Techniques in Human Communication Research. (Academic Press; pp. 552) New York and London, pp. 259-308

716. KRISHNAIAH, P. R. (General Ed.; 1980/88): Handbook of Statistics. Vol. 1-14, Vol. 1: Analysis of Variance (716.), Vol. 2: Classification, Pattern Recognition and Reduction of Dimensionality (717.), Vol. 3: Time Series in the Frequency Domain (718.), Vol. 4: Nonparametric Methods through (719.), Vol. 5: Time Series in the Time Domain (720.), Vol. 6: Sampling bis (721.), Vol. 7: Quality Control and Reliability (722.), Vol. 8: Statistical 723. Models in Biological and Medical Sciences (723.), Vol. 9-14: (not yet titled). (North-Holland; pp. 1002, 904, 482, 936, 496) New York and Amsterdam

724. KRITZER, H. M. (1977): Analyzing measures of association derived from contingency tables. Sociological Methods and Research **5**, 387-418 [see also **8** (1980), 420-433]

725. KROON, DE, J. and P. VAN DER LAAN (1983): A generalization of Friedman's rank statistic. STNE **37**, 1-14

726. KRUG, W. (1976): Quantifizierung des systematischen Fehlers in wirtschafts- und sozialstatistischen Daten. Dargestellt an der Statistik der Erwerbstätigkeit. (Duncker u. Humblot, 109 S.) Berlin

727. KRUG, W. und M. NOURNEY (1982): Wirtschafts- und Sozialstatistik: Gewinnung von Daten. (Oldenbourg; 214 S.) München, Wien

728. KRUSKAL, W. H. (1982): Criteria for judging statistical graphics. Utilitas Mathematica **21B**, 283-310

729. KRUSKAL, W. and F. MOSTELLER (1979/80): Representative Sampling, I-IV. INTR **47**, 13-24, 111-127, 245-265; **48**, 169-195

730. Krzanowski, W.J. (1980): Mixtures of continuous and categorical variables in discriminant analysis. BICS **36**, 493–499

731. Krzysko, M. (1982): Canonical analysis. BIJL **24**, 211–228

732. Kshirsagar, A.M. (1983): A Course in Linear Models. (M. Dekker; pp. 448) New York

733. Ku, H.H. (Ed.; 1969): Precision Measurement and Calibration. Selected NBS Papers on Statistical Concepts and Procedures. (US Dept. Commerce, National Bureau of Standards; pp. 436) Washington, D.C.

734. Kudo, A. and J.S. Yao (1982): Tables for testing ordered alternatives in an analysis of variance without replications. BIKA **69**, 237–238

735. Kübler, H. (1979): On the parameter of the three-parameter distributions: lognormal, gamma and Weibull. Statistische Hefte **20**, 68–125

736. Küffner, H. und R. Wittenberg (1985): Datenanalyse für statistische Auswertungen. Eine Einführung in SPSS, BMDP und SAS. (G. Fischer; 289 S.) Stuttgart und New York

737. Kugel, P. (1977): Induction, pure and simple. Information and Control **35**, 276–336

738. Kupper, L.L., J.M. Karon, D.G. Kleinbaum, H. Morgenstern and D.K. Lewis (1981): Matching in epidemiologic studies: validity and efficiency considerations. BICS **37**, 271–291

739. Kurtz, A.K. and S.T. Mayo (1979): Statistical Methods in Education and Psychology. (Springer; pp. 538) New York

740. Lachenbruch, P.A. (1975): Discriminant Analysis. (Hafner Press; pp. 128) New York

741. Lachenbruch, P.A. (1983): Statistical programs for microcomputers. Byte **8**, 460–570

742. Lachenbruch, P.A. and W.R. Clarke (1980): Discriminant analysis and its applications in epidemiology. MIME **19**, 220–226

743. Lachenbruch, P.A. and M. Goldstein (1979): Discriminant analysis. BICS **35**, 69–85

744. Lackritz, J.R. (1984): Exact p values for F and t tests. AMST **38**, 312–314

745. Lahey, A.L., R.G. Downey and F.E. Saal (1983): Intraclass correlations: there's more than meets the eye. PYBU **93**, 586–595

746. Laisiepen, K., E. Lutterbeck und K.-H. Meyer-Uhlenried (1980): Grundlagen der praktischen Information und Dokumentation. 2. völlig neubearb. Aufl. (Saur; 826 S.) München

747. Lam, H.K. (1982): Bounded multinomial distribution. CSTH **11**, 1869–1880

748. Lam, H.K. and Y.-F. Ho (1985): On the bounded binomial distribution and its parameter estimation. CSSM **14**, 43–53

749. Lancaster, P. and M. Tismenetsky (1985): The Theory of Matrices. With Applications. 2nd ed. (Academic Press; pp. 592) London

750. Land, K.C. and A. Rogers (Eds.; 1982): Multidimensional Mathematical Demography. (Academic Press; pp. 605) New York

751. LANE, P. W. and J. A. NELDER (1982): Analysis of covariance and standardization as instances of prediction. BICS **38**, 613–621

752. LANGEHEINE, R. (1980): Log-lineare Modelle zur multivariaten Analyse qualitativer Daten. Eine Einführung. (Oldenbourg; 124 S.) München, Wien

753. LAST, J. M. (Ed.; 1983): A Dictionary of Epidemiology. (IEA/WHO) (Oxford Univ. Press; pp. 114) New York, Oxford, Toronto

754. LAUDAN, L. (1977): Progress and Its Problems. Toward a Theory of Scientific Growth. (Univ. of Calif. Press; pp. 257) Berkeley

755. LAUNER, R. L. and A. F. SIEGEL (Eds.; 1982): Modern Data Analysis. (Academic Press; pp. 201) New York

756. LAUNER, R. L. and G. N. WILKINSON (Eds.; 1979): Robustness in Statistics. (Academic Press; pp. 320) New York

757. LAUX, H. (1982): Entscheidungstheorie. Grundlagen. (Springer; 349 S.) Berlin, Heidelberg, New York

758. LAUX, H. (1982): Entscheidungstheorie II. Erweiterung und Vertiefung. (Springer; 280 S.) Berlin, Heidelberg, New York

759. LAW, A. M. (1983): Statistical analysis of simulation output data. Operations Research **31**, 983–1029 (see also the other 8 papers of this special issue, pp. 1030–1199)

760. LAW, A. M. and W. D. KELTON (1982): Simulation Modeling and Analysis. (McGraw-Hill; pp. 400) New York

761. LAWLESS, J. F. (1982): Statistical Models and Methods for Lifetime Data. (Wiley-Interscience; pp. 580) New York

762. LAWLESS, J. F. (1983): Statistical methods in reliability. With discussion. TECS **25**, 305–335

763. LAWRENCE, R. J. (1984): The lognormal distribution of the duration of strikes. JRSA **147**, 464–483

764. LAWSON, A. (1983): Rank analysis of covariance: alternative approaches. STAN **32**, 331–337

765. LAWSON, J. S. (1982): Application of robust regression in designed industrial experiments. JQTE **14**, 19–33

766. LEACH, C. (1979): Introduction to Statistics. A nonparametric Approach for the social sciences. (Wiley; pp. 339) New York

767. LEBART, L., A. MORINEAU and K. M. WARWICK (1984): Multivariate Descriptive Statistical Analysis. Correspondence Analysis and Related Techniques for Large Matrices. (Wiley; pp. 231) New York

768. LEDERMANN, W. and Emlyn LLOYD (Eds.; 1980): Handbook of Applicable Mathematics. Vol. II: Probability. (Wiley; pp. 450) Chichester and New York

769. LEDERMANN, W. and EMLYN LLOYD (Eds.; 1984): Handbook of Applicable Mathematics. Vol. VI: Statistics. Parts A and B (Wiley; pp. 498 + Appendix + Index and pp. 942 + Appendix + Index) Chichester and New York

770. LEE, Elisa T. (1980): Statistical Methods for Survival Data Analysis. (Lifetime Learning Publications; pp. 557) Belmont, Calif. 94002

771. LEE, Y. S. (1972): Tables of upper percentage points of the multiple correlation coefficient. BIKA **59**, 175-189

772. LEHMACHER, W. (1979): A new nonparametric approach to the comparison of *k* independent samples of response curves II: a *k* sample generalization of the Friedman test. BIJL **21**, 123-130

773. LEHMACHER, W. (1980): Simultaneous sign tests for marginal homogeneity of square contingency tables. BIJL **22**, 795-798

774. LEHMANN, E. L. (1983): Theory of Point Estimation. (Wiley; pp. 506) New York

775. LEHMANN, R. (1977): General derivation of partial and multiple rank correlation coefficients. BIJL **19**, 229-236

776. LEIBBRAND, D. (1984): Datenhaltung und Erfassung bei klinischen Studien. EDVM **15**, 33-40

777. LEINER, B. (1978): Spektralanalyse ökonomischer Zeitreihen. Einführung in Theorie und Praxis moderner Zeitreihenanalyse. 2. erw. Aufl. (Gabler; 144 S.) Wiesbaden

778. LEINER, B. (1982): Einführung in die Zeitreihenanalyse. (Oldenbourg; 141 S.) München und Wien

779. LEINER, B. (1985): Stichprobentheorie. Grundlagen, Theorie und Technik. (R. Oldenbourg; 156 S.) München und Wien

780. LEMESHOW, S. and D. W. HOSMER (1981): A comparison of sample size determination methods in the two group trial where the underlying disease is rare. CSSM B **10**, 437-449

781. LENZ, H.-J., G. B. WETHERILL and P. Th. WILRICH (Eds.; 1981, 1984) Frontiers in Statistical Quality Control. Vol. 1 and 2. (Physica-Vlg.; pp. 288 and 292) Würzburg and Wien

782. LEPAGE, Y. (1977): A class of nonparametric tests for location and scale parameters. CSTH A **6**, 649-658

783. LESAFFRE, E. (1983): Normality tests and transformations. Pattern Recognition Letters **1**, 187-199

784. LEVENBACH, H. and J. P. CLEARY (1981): The Beginning Forecaster: The Forecasting Process through Data Analysis. (Lifetime Learning Publications; pp. 372) Belmont, Calif.

785. LEVENBACH, H. and J. P. CLEARY (1984): The Modern Forecasting Process Through Data Analysis. (Lifetime Learning Publications; pp. 450) Belmont, Calif. 94002

786. LEVINE, D. M., M. L. BERENSON and D. STEPHAN (1983): (1) Using the SPSS Batch System with Basic Business Statistics. (2) Using the Statistical Analysis System (SAS) with Basic Business Statistics. (Prentice-Hall; pp. 69 and 83) Englewood Cliffs, New Jersey

787. LEVY, K. J. (1980): Nonparametric applications of Shaffer's extension of Dunnett's procedure. AMST **34**, 99-102

173

788. LEWIS, J. A. (1983): Clinical trials: statistical developments of practical benefit to the pharmaceutical industry. With discussion. JRSA **146**, 362–393

789. LI, C. C. (1982): Analysis of Unbalanced Data. A Pre-Program Introduction. (Cambridge Univ. Press; pp. 145) Cambridge, London, New York

790. LI, L. and W. R. SCHUCANY (1975): Some Properties of a test for concordance of two groups of rankings. BIKA **62**, 417–423

791. LIDDELL, F. D. K., J. C. MCDONALD and D. C. THOMAS (1977): Methods of cohort analysis: Appraisal by application to asbestos mining. With discussion JRSA **140**, 469–491

792. LIEBETRAU, A. M. (1983): Measures of Association. (Sage Publicat. Series 07-032; pp. 94) Beverley Hills and London

793. LIENERT, G. A. (1973/78): Verteilungsfreie Methoden in der Biostatistik. 2. Aufl. (Bd. I und II; Tafelband) (A. Hain; 736 S., 1246 S., 686 S.) Meisenheim am Glan 1973, 1978, 1975

794. LIENERT, G. A., O. LUDWIG and K. ROCKENFELLER (1982): Tables of the critical values for simultaneous and sequential Bonferroni-z-tests. BIJL **24**, 239–255

795. LIENERT, G. A. and L. A. MARASCUILO (1980): Comparing treatment-induced changes for k independent samples of paired observations. BIJL **22**, 762–777

796. LIKEŠ, J. and J. LAGA (1980): Probabilities $P(S \geqq s)$ for the Friedman statistic. BIJL **22**, 433–440

797. LILIENFELD, A. M. and D. E. LILIENFELD (1980): Foundation of Epidemiology. 2nd rev. ed. (Oxford Univ. Press; pp. 400) New York

798. LINDEMAN, R. H., P. F. MERENDA and Ruth Z. GOLD (1980): Introduction to Bivariate and Multivariate Analysis. (Scott, Foreman and Company; pp. 444) Glenview, Illinois; Dallas, Texas; …, and London

799. LINDER, A. und W. BERCHTOLD (1976): Statistische Auswertung von Prozentzahlen. Probit- und Logitanalyse mit EDV. (Uni-Tb. Nr. 522) (Birkhäuser; 232 S.) Basel und Stuttgart

800. LINDER, A. und W. BERCHTOLD (1982): Statistische Methoden III: Multivariate Verfahren. (UTB 1189) (Birkhäuser; 218 S.) Basel, Boston, Stuttgart

801. LINDLEY, D. V. (1978): The Bayesian approach. Scand. J. Statist. **5**, 1–26

802. LINDLEY, D. V., D. A. EAST and P. A. HAMILTON (1960): Tables for making inferences about the variance of a normal distribution. BIKA **47**, 433–437

803. LINKE, W. (1983): Drei Verfahren zur Vorausschätzung der Privathaushalte. Zeitschrift für Bevölkerungswissenschaft **9**, 27–46

804. LITTLE, R. J. A. (1982): Models for nonresponse in sample surveys. JASA **77**, 237–250

805. LITTLE, R. J. A. and T. W. PULLUM (1979): The general linear model and direct standardization. A comparison. Sociological Methods and Research **7**, 475–501

806. LONG, J.S. (1983): Confirmatory Factor Analysis: A Preface to LISREL. (Sage Publicat. Series 07-033; pp.88) Beverly Hills and London

807. LONG, J.S. (1983): Covariance Structure Models: An Introduction to LISREL. (Sage Publicat. Series 07-034; pp.93) Beverly Hills and London

808. LOONEY, S.W. and T.R.GULLEDGE, Jr. (1985): Use of the correlation coefficient with normal probability plots. AMST **39**, 75–79 [cf. 236]

809. LORENZI, R. (1981): Korrespondenzanalyse. Statistische Hefte **22**, 176–194

810. LORR, M. (1983): Cluster Analysis for Social Scientists. Techniques for Analyzing and Simplifying Complex Blocks of Data. (Jossey-Bass; pp.233) San Francisco, Washington, London

811. LOUV, W.C. (1984): Adaptive filtering. TECS **26**, 399–409

812. LOYER, M. (1983): Bad probability, good statistics, and group testing for binomial estimation. AMST **37**, 57–59

813. LUBIN, J.H. (1981): An empirical evaluation of the use of conditional and unconditional likelihoods for case-control data. BIKA **68**, 567–571

814. LUNNEBORG, C.E. and R.D.ABBOTT (1983): Elementary Multivariate Analysis for the Behavioral Sciences. Applications of Basic Structure. (Elsevier/North-Holland; pp.522) New York, Amsterdam, Oxford

815. LWIN, T. and J.S.MARITZ (1982): An analysis of the linear-calibration controversy from the perspective of compound estimation. TECS **24**, 235–242

816. MAASS, S. (1983): Statistik für Wirtschafts- und Sozialwissenschaftler I. Wahrscheinlichkeitstheorie. (Heidelberger TB, Bd.232) (Springer; 403 S.) Berlin, Heidelberg, New York, Tokyo

817. MAASS, S., H.MÜRDTER und H.Ch.RIESS (1983): Statistik für Wirtschafts- und Sozialwissenschaftler II. Induktive Statistik. (Heidelberger TB, Bd.233) (Springer; 360 S.) Berlin, Heidelberg, New York, Tokyo

818. MACDONALD, J.S. and Leatrice MACDONALD (1982): The Demography of Ageing. (Croom Helm; pp.320) London

819. MADHAVA RAO, K.S. and A.P.GORE (1982): Nonparametric tests for intercept in linear regression problems. Australian Journal of Statistics **24**, 42–50

820. MADOW, W.G., H.NISSELSON, I.OLKIN and D.B.RUBIN (Eds.; 1984): Incomplete Data in Sample Surveys. Vol.1: Report and Case Studies, Vol.2: Theory and Bibliographies, Vol.3: Proceedings of the Symposium. (Academic Press; pp.512, 608, 440) New York

821. MAGE, D.T. (1982): An objective graphical method for testing normal distributional assumptions using probability plots. AMST **36**, 116–120

822. MAGIDSON, J. (1982): Some common pitfalls in causal analysis of categorical data. Journal of Marketing Research **19**, 461–471

823. MAIBAUM, G. (1980): Wahrscheinlichkeitsrechnung. 3.Aufl. (H.Deutsch; 223 S.) Thun und Frankfurt/M.

824. MAINDONALD, J.H. (1984): Statistical Computation. (Wiley; pp. 370) New York

825. MAKRIDAKIS, S. and M. HIBON (1979): Accuracy of forecasting. An empirical investigation. With discussion. JRSA **142**, 97-145

826. MAKRIDAKIS, S., S.C. WHEELWRIGHT and V.E. McGEE (1983): Forecasting. Methods and Applications. 2nd ed. (Wiley; pp. 944) New York

827. MALINOWSKI, E.R. and D.G. HOWE (1981): Factor Analysis in Chemistry. (Wiley; pp. 251) New York

828. MALLOWS, C.L. (1979): Robust methods – some examples of their use. AMST **33**, 179-184

829. MALLOWS, C.L. and J.W. TUKEY (1982): An Overview of Techniques of Data Analysis, Emphasizing its Exploratory Aspects. In: Tiago di Oliveira, J. and B. Epstein (Eds.): Some Recent Advances in Statistics. (Academic Press; pp. 248) London and New York, pp. 111-172

830. MALLOWS, C.L. and P. WALLEY (1980): A theory of data-analysis? Proceedings of the Business and Economics Statistics Section. American Statistical Association, 8-14

831. MANDEL, J. (1982): Use of the singular value decomposition in regression analysis. AMST **36**, 15-24 [see also **39** (1985), 63-66]

832. MANDEL, J. (1984): Fitting straight lines when both variables are subject to error. JQTE **16**, 1-14

833. MANSKI, C.F. and D. McFADDEN (Eds.; 1981): Structural Analysis of Discrete Data: With Econometric Applications. (The MIT Press; pp. 477) Cambridge, Mass.

834. MANTEL, N. (1967): Ranking procedures for arbitrarily restricted observations. BICS **23**, 65-78

835. MARASCUILO, L.A. and J.R. LEVIN (1983): Multivariate Statistics in the Social Sciences – A Researcher's Guide. (Brooks/Cole Publ. Comp.; pp. 530) Monterrey, Calif.

836. MARASCUILO, L.A. and Maryellen McSWEENEY (1977): Nonparametric and Distribution-Free Methods for the Social Sciences. (Brooks/Cole; pp. 556) Monterey, Calif.

837. MARASCUILO, L.A. and R.C. SERLIN (1979): Tests and contrasts for comparing change parameters for a multiple McNemar data model. British Journal of Mathematical and Statistical Psychology **32**, 105-112

838. MARCUS, Ruth (1981): Simultaneous confidence intervals for monotone contrasts of normal means. Australian Journal of Statistics **23**, 214-223

839. MARDIA, K.V. (1972): Statistics of Directional Data. (Acad. Press; pp. 355) London

840. MARDIA, K.V., J.T. KENT and J.M. BIBBY (1979): Multivariate Analysis. (Academic Press; pp. 521) London

841. MARDIA, K.V. and P.J. ZENROCH (1978): Tables of the *F*- and Related Distributions with Algorithms. (Academic Press; pp. 286) New York

842. MARKUS, G.B. (1979): Analyzing Panel Data (Sage Publicat. Series 07-018; pp. 71) Beverley Hills and London

176

843. MARQUARDT, D. W. and R. D. SNEE (1975): Ridge regression in practice. AMST **29**, 3–20

844. MARRIOTT, F. H. C. (1982): Optimization methods of cluster analysis. BIKA **69**, 417–421

845. MARSH, Catherine (1982): The Survey Method. The Contribution of Surveys to Sociological Explanation. (Allen and Unwin; pp. 180) London

846. MARTIN, Margaret E. (1981): Statistical practice in bureaucracies. JASA **76**, 1–8

847. MARTZ, H. F. and R. A. WALLER (1982): Bayesian Reliability Analysis. (Wiley; pp. 745) New York

848. MARUYAMA, G. and B. McGARNEY (1980): Evaluating causal models: an application of maximum-likelihood analysis of structural equations. PYBU **87**, 502–512

849. MARYANSKI, F. J. (1980): Digital Computer Simulation. (Hayden, pp. 336) Rochelle Park, NJ 07662

850. MASING, W. (Hrsg.; 1980): Handbuch der Qualitätssicherung. (Hanser; 970 S.) München

851. MASON, W. M. and S. E. FIENBERG (Eds.; 1984): Cohort Analysis in Social Research: Beyond the Identification Problem. (Springer; pp. 400) Berlin, Heidelberg, New York, Tokyo

852. MATHER, K. and J. L. JINKS (1982): Biometrical Genetics. The Study of Continuous Variation. 3rd ed. (Chapman and Hall; pp. 396) New York

853. MATTHEWS, J. N. S. (1984): Robust methods in the assessment of multivariate normality. APST **33**, 272–277

854. MAXWELL, A. E. (1976): Analysis of contingency tables and further reasons for not using Yates correction in 2×2 tables. Canad. J. Statist. C **4**, 277–290

855. MAXWELL, A. E. (1977): Multivariate Analysis in Behavioral Research. (Chapman and Hall; pp. 164) London

856. MAXWELL, S. E. (1980): Pairwise multiple comparisons in repeated measures designs. EDUC **5**, 269–287

857. MAXWELL, S. E., H. D. DELANEY and C. A. DILL (1984): Another look at ANCOVA versus blocking. PYBU **95**, 136–147

858. McCABE, G. P. (1984): Principal variables. TECS **26**, 136–144

859. McCLEARY, R. and R. A. HAY, Jr. (1980): Applied Time Series Analysis for the Social Sciences. (Stage Publications; pp. 330) Beverly Hills, Calif.

860. McCORMICK, G. P. (1983): Nonlinear Programming: Theory, Algorithms, and Applications. (Wiley; pp. 444) New York

861. McCULLAGH, P. (1978): A class of parametric models for the analysis of square contingency tables with ordered categories. BIKA **65**, 413–418

862. McCULLAGH, P. (1980): Regression models for ordinal data. With discussion. JRSB **42**, 109–142

863. McCULLAGH, P. (1983): Quasi-likelihood functions. Annals of Statistics **11**, 59–67

864. McCullagh, P. and J. A. Nelder (1983): Generalized Linear Models. (Chapman and Hall; pp. 280) London

865. McDowall, D., R. McCleary, E. E. Meidinger and R. A. Hay, Jr. (1980): Interrupted Time Series Analysis. (Stage Publicat. Series 07-021; pp. 95) Beverly Hills and London

866. McGill, R., J. W. Tukey and W. A. Larsen (1978): Variations of box plots. AMST 32, 12–16

867. McIntyre, S. H. and A. B. Ryans (1983): Task effects on decision quality in traveling salesperson problems. Organizational Behavior and Human Performance 32, 344–369

868. McKinlay, Sonja M. (1977): Pair-matching. A reappraisal of a popular technique. BICS 33, 725–735

869. McLachlan, G. J. and S. Ganesalingam (1982): Updating a discriminant function on the basis of unclassified data. CSSM 11, 753–767

870. McLaughlin, M. L. (1980): Discriminant Analysis in Communication Research. Chapter 6 in P. R. Monge and J. N. Cappella (Eds.): Multivariate Techniques in Human Communication Research. (Academic Press; pp. 552) New York and London, pp. 175–204

871. McNeil, D. R. (1977): Interactive Data Analysis. (Wiley; pp. 186) New York and London

872. Mczynski, M. J. and P. K. Pathak (1980): Integration of surveys. Scand. J. Statist. 7, 130–138

873. Mead, R. and D. J. Pike (1975): A review of response surface methodology from a biometric viewpoint. BICS 31, 803–851

874. Medhi, J. (1982): Stochastic Processes. (Wiley Eastern Limited, pp. 387) New Delhi

875. Mee, R. W. and D. B. Owen (1983): A simple approximation for bivariate normal probabilities. JQTE 15, 72–75

876. Meeker, W. Q. and G. J. Hahn (1982): Sample sizes for prediction intervals. JQTE 14, 201–206

877. Mehta, C. R. and N. R. Patel (1983): A network algorithm for performing Fisher's exact test in $r \times c$ contingency tables. JASA 78, 427–434

878. Meinert, C. L. (1985): Clinical Trials: Design, Conduct and Analysis. (Oxford Univ. Press; pp. 512) New York

879. Menges, G. (1980): Adaptive Statistik. Statistische Hefte 21, 182–208

880. Menges, G. (1982): Die Statistik. Zwölf Stationen des statistischen Arbeitens. (Dr. Th. Gabler; 505 S.) Wiesbaden

881. Mertens, P. (Hrsg.; 1981): Prognoserechnung. 4. erweit. Aufl. (Physica-Verlag; 364 S.) Würzburg

882. Meschkowski, H. (1984): Was wir wirklich wissen. Die exakten Wissenschaften und ihr Beitrag zur Erkenntnis. (Piper; 309 S.) München u. Zürich

883. Messerschmitt-Bölkow-Blohm GmbH (Hrsg.; 1977): Technische Zuverlässigkeit. Problematik, Mathematische Grundlagen, Untersuchungs-

methoden, Anwendungen. 2. neubearb. Aufl. (Springer; 308 S.) Heidelberg

884. MESSICK, D. M. und J. P. VAN DE GEER (1981): A reversal paradox. PYBU **90**, 582-593

885. MEYER, M. (1983): Operations Research. Systemforschung. Eine Einführung in die praktische Bedeutung. (Uni-Tb. 1231) (Fischer; 202 S.) Stuttgart

886. MICHAEL, J. R. (1983): The stabilized probability plot. BIKA **70**, 11-17

887. MIETTINNEN, O. S. and E. F. COOK (1981): Confounding: essence and detection. AMEP **114**, 593-603

888. MIKÉ, V. and K. E. STANLEY (Eds.; 1982): Statistics in Medical Research. Methods and Issues with Applications in Cancer Research. (Wiley; pp. 551) New York

889. MILES, M. B. and A. M. HUBERMAN (1984): Qualitative Data Analysis. (Sage; pp. 256) Beverly Hills and London

890. MILLER, A. J. (1984): Selection of subsets of regression variables. With discussion. JRSA **147**, 389-425

891. MILLER, A. R. (1981): BASIC Programs for Scientists and Engineers. (SYBEX; pp. 318) Berkeley, Paris, Düsseldorf

892. MILLER, K. S. (1980): Hypothesis Testing with Complex Distributions. (R. E. Krieger; pp. 184) Huntington, N. Y.

893. MILLER, R. (1981): Survival Analysis. (Wiley-Interscience; pp. 238) New York

894. MILLER, R. E. and P. D. BLAIR (1985): Input-Output Analysis. Foundations and Extensions. (Prentice-Hall; pp. 448) Englewood Cliffs, N. J.

895. MILLER, R. G., Jr. (1981): Simultaneous Statistical Inference. 2nd ed. (Springer; pp. 299) New York, Heidelberg, Berlin

896. MILLIKEN, G. A. and E. JOHNSON (1984): Analysis of Messy Data: Vol. I – Designed Experiments. (Lifetime Learning Publicat.; pp. 600) Belmont, Calif. 94002

897. MILTON, J. S. and J. O. TSOKOS (1983): Statistical Methods in the Biological and Health Sciences. (McGraw-Hill; pp. 512) New York

898. MOHN, E. (1979): Confidence estimation of measures of location in the log normal distribution. BIKA **66**, 567-575

899. MONGE, P. R. (1980): Multivariate Multiple Regression. Chapter 2 in P. R. Monge and J. N. Cappella (Eds.): Multivariate Techniques in Human Communication Research. (Academic Press; pp. 552) New York and London, pp. 13-56

900. MONGE, P. R. and J. N. CAPPELLA (Eds.; 1980): Multivariate Techniques in Human Communication Research. (Academic Press; pp. 552) New York and London

901. MONLEZUN, C. J., D. C. BLOUIN and Linda C. MALONE (1984): Contrasting split plot and repeated measures experiments and analyses. With comments. AMST **38**, 21-31 (see also 331-332)

902. MONTGOMERY, D.C. and Elizabeth A. PECK (1982): Introduction to Linear Regression Analysis. (Wiley-Interscience; pp. 504) New York

903. MOORE, P.G. (1983): The Business of Risk. (Cambridge Univ. Press; pp. 244) Cambridge

904. MOORE, S.A. (1981): Calculating probabilities using recurrence relations. Teaching Statistics (Sheffied) **3**, 43–47

905. MORAN, P.A.P. (1980): Calculation of the normal distribution function. BIKA **67**, 675–677

906. MORETTIN, P.A. (1984): The Levinson algorithm and its applications in time series analysis. INTR **52**, 83–92

907. MORGAN, B.J.T. (1981): Three applications of methods of cluster-analysis. STAN **30**, 205–223

908. MORGAN, B.J.T. (1984): Elements of Simulation. (Chapman and Hall; pp. 230) London

909. MORGAN, B.J.T. and P.M. NORTH (Eds.; 1985): Statistics in Ornithology. (Lecture Notes in Statistics, Vol. 29) (Springer; pp. 418) New York, Berlin, Heidelberg, Tokyo

910. MORRISON, A.S. (1985): Screening in Chronic Disease. (Oxford Univ. Press; pp. 182) New York

911. MORRISON, D.F. (1983): Applied Linear Statistical Methods. (Prentice-Hall; pp. 562) Englewood Cliffs, N.J.

912. MORTON, R.H. (1983): Response surface methodology. Math. Scientist **8**, 31–52

913. MOSER, C. (1980): Statistics and public policy. JRSA **143**, 1–28

914. MOSES, L.E. (1978): Charts for finding upper percentage points of Student's t in the range .01 to .00001. CSSM B **7**, 479–490

915. MOSTELLER, F. and J.W. TUKEY (1968): Data Analysis, Including Statistics. In: Handbook of Social Psychology. G. Lindzey and E. Aronson (Eds.), 2nd ed., Chapter 10, pp. 80/203. (Addison-Wesley) Reading, Mass.

916. MOSTELLER, F. and J.W. TUKEY (1977): Data Analysis and Regression. A Second Course in Statistics. (Addison-Wesley; pp. 588) Reading, Mass.

917. MOULD, R.F. (1983): Cancer Statistics. (A. Hilger; pp. 210) Bristol

918. MÜLLER, G.W. und T. KICK (1983): BASIC-Programme für die angewandte Statistik. 32 Programme für Kleincomputer. (Oldenbourg; 305 S.) München und Wien

919. MÜLLER, H.G. (1985): Nichtparametrische Regression für die Analyse von Verlaufskurven. In: Pflug, G. Ch. (Hrsg.; 1985): Neuere Verfahren der nichtparametrischen Statistik. (Mediz. Informatik und Statist. Bd. 60) (Springer; 129 S.) Berlin, Heidelberg, New York, Tokyo; S. 88–108

920. MÜLLER, P.H. (Hrsg.; 1980): Lexikon der Stochastik. Wahrscheinlichkeitsrechnung und Mathematische Statistik. 3. ber. Aufl. (Akademie Vgl.; 445 S.) [4. unv. Aufl. 1983]

921. MÜLLER, P.H., P.NEUMANN und Regina STORM (1977): Tafeln der mathematischen Statistik. 2. verb. Aufl. (C.Hanser; 275 S.) München und Wien

922. MULLER, K.E. (1982): Understanding canonical correlation through the general linear model and principal components. AMST **36**, 342-354

923. MUNDEL, A.B. (1984): Group testing. JQTE **16**, 181-188

924. MURDOCH, J. (1979): Control Charts. (Macmillan; pp.150) New York

925. MURPHY, B.P. (1985): The microcomputer as the statistician's mainframe. SSNL **11**, 95-99

926. MUTHÉN, Linda (1984): BMDP distributes a new structural equation program. SSNL **10**, 37

927. NAMBOODIRI, K. (1984): Matrix Algebra. An Introduction. (Sage Publicat. Series 07-038; pp.96) Beverly Hills and London

928. NAMBOODIRI, N.K. (Ed.; 1978): Survey Sampling and Measurement. (Academic Press; pp.364) New York and London

929. NAROLA, S.C. and J.F.WELLINGTON (1982): The minimum sum of absolute errors regression: a state of the art survey. INTR **50,** 317-326

930. NARULA, S.C. (1978): Orthogonal polynomial regression for unequal spacing and frequencies. JQTE **10**, 170-179

931. NARULA, S.C. (1979): Orthogonal polynomial regression. INTR **47,** 31-36

932. NAUS, J.I. (1975): Data Quality Control and Editing. (M.Dekker; pp.216) New York

933. NEFFENDORF, H. (1983): Statistical packages for microcomputers. A listing. AMST **37**, 83-86

934. NEILL, J.J. and O.J.DUNN (1975): Equality of dependent correlation coefficients. BICS **31**, 531-543

935. NEILL, J.W. and D.E.JOHNSON (1984): Testing for lack of fit in regression - a review. CSTH **13**, 485-511

936. NELDER, J.A. (1980): Iterative weighted least squares, an algorithm for many occasions. In E.Diday et al. (Eds.): Data Analysis and Informatics. (North-Holland Publ. Comp.) Amsterdam/New York, pp.75-81

937. NELDER, J.A. (1982): Linear models and non-orthogonal data. Utilitas Mathematica **21B**, 141-152

938. NELSON, L.S. (1977): Tables for testing ordered alternatives in an analysis of variance. BIKA **64**, 335-338

939. NELSON, L.S. (1979): Too many defectives in too short a time? JQTE **11,** 160-161

940. NELSON, L.S. (1982): Extreme screening designs. JQTE **14**, 99-100

941. NELSON, L.S. (1983): Exact critical values for the analysis of means. JQTE **15**, 40-44

942. NELSON, L.S. (1983): The deceptiveness of moving averages. JQTE **15,** 99-100

943. NELSON, L. S. (1983): Expected normal scores, a useful transformation. JQTE **15**, 144–146

944. NELSON, P. R. (1983): A comparison of sample sizes for the analysis of means and the analysis of variance. JQTE **15**, 33–39

945. NELSON, P. R. (1983): The analysis of means for balanced experimental designs. (Computer Program). JQTE **15**, 45–54

946. NELSON, P. R. (1985): Power curves for the analysis of means. TECS **27**, 65–73

947. NELSON, T. O. (1984): A comparison of current measures of the accuracy of feeling-of-knowing predictions. PYBU **95**, 109–133

948. NELSON, W. (1970): Confidence intervals for the ratio of two Poisson means and Poisson predictor intervals. IEEE Transactions on Reliability R-**19**. No. 2, 42–49

949. NELSON, W. (1970): A statistical prediction interval for availability. IEEE Transactions on Reliability R-**19**, No. 4, 179–182

950. NELSON, W. (1972): Theory and applications of hazard plotting for censored failure data. TECS **14**, 945–966

951. NELSON, W. (1978): Life data analysis for units inspected once for failure (quantal response data). IEEE Transactions on Reliability R-**27**, No. 4, 274–279

952. NELSON, W. (1982): Applied Life Data Analysis. (Wiley; pp. 634) New York

953. NERLOVE, M., D. M. GRETHER and J. L. CARVALHO (1979): Analysis of Economic Time Series. A. Synthesis. (Academic Press; pp. 468) New York

954. NESSELROADE, J. R. and P. B. BALTES (Eds.; 1979): Longitudinal Research in the Study of Behavior and Development. (Academic Press; pp. 386) New York, London, Toronto

955. NETER, J. and W. WASSERMAN (1974): Applied Linear Statistical Models. Regression, Analysis of Variance and Experimental Design. (R. D. Irwin; pp. 842). Homewood, Ill.

956. NETER, J., W. WASSERMAN and M. H. KUTNER (1983): Applied Linear Regression Models. (R. D. Irwin; pp. 547) Homewood, Ill.

957. NEUHAUS, G. and E. KREMER (1981): Repeated chi-square testing. CSSM B **10**, 143–161

958. NEWBOLD, P. (1984): Some recent developments in time series analysis – II. INTR **52**, 183–192

959. NEWELL, D. (1982): The role of the statistician as an expert witness. With discussion. JRSA **145**, 403–409, 426–438

960. NICHOLS, W. G. and J. D. GIBBONS (1979): Parameter measures of skewness. CSSM B **8**, 161–167

961. NICOLAS, M. (1948): Ein unentbehrlicher Mittelwert. Statistische Praxis **3**, 185–186

962. NIEMANN, H. (1979): Mustererkennung – Einführung und Übersicht. Informatik-Spektrum **2**, 12–24

963. NIEMANN, H. (1980): Mustererkennung – Anwendungen. Informatik-Spektrum **3**, 19–30

964. NIEMANN, H. (1983): Klassifikation von Mustern. (Springer; 340 S.) Heidelberg

965. NISHISATO, S. (1980): Analysis of Categorical Data: Dual Scaling and Its Applications. (Math. Expos. No. 24) (Univ. of Toronto Press; pp. 276) Toronto

966. NÖBAUER, W. und W. TIMISCHL (1979): Mathematische Modelle in der Biologie. (Vieweg; 232 S.) Braunschweig und Wiesbaden

967. NOELLE-NEUMANN, Elisabeth und E. PIEL (Hrsg.; 1983): Eine Generation später. Bundesrepublik Deutschland 1953–1979. Institut für Demoskopie Allensbach. (Saur; 272 S.) München, New York, London, Paris

968. NORCLIFFE, G. B. (1981): Statistik für Geographen. Eine Einführung. Übers. v. T. Thyssen. (Springer; 250 S.) Berlin, Heidelberg, New York

969. NORTON, R. W. (1980): Nonmetric Multidimensional Scaling in Communication Research: Smallest Space Analysis. Chapter 10 in P. R. Monge and J. N. Cappella (Eds.): Multivariate Techniques in Human Communication Research. (Academic Press; pp. 552) New York and London, pp. 309–331

970. NORTON, V. (1983): A simple algorithm for computing the non-central F-distribution. APST **32**, 84–85

971. NORUSIS, Marija (1983): SPSS-X Introductory Statistics Guide. (McGraw-Hill; pp. 276) New York

972. NORUSIS, Marija (1985): SPSS-X Advanced Statistics Guide. (McGraw-Hill; pp. 320) New York

973. NOWAK, W. (1980): Das Statistische Bundesamt. Ein Überblick über Aufgaben und Organisation. Wirtschaftswissenschaftliches Studium **9**, 238–241

974. OAKES, D. (1983): Survival analysis. European Journal of Operational Research **12**, 3–14

975. O'BRIEN, R. G. and Mary KISTER KAISER (1985): MANOVA method for analyzing repeated measures designs: an extensive primer. PYBU **97**, 316–333

976. ODEH, R. E. (1971): On Jonckheere's k-sample test against ordered alternatives. TECS **13**, 912–918

977. ODEH, R. E. (1972): On the power of Jonckheere's k-sample test against ordered alternatives. BIKA **59**, 467–471

978. ODEH, R. E. (1977): Extended tables of the distribution of Friedman's S-statistic in the two-way layout. CSSM B **6**, 29–48

979. ODEH, R. E. (1977): The exact distribution of Page's L-statistic in the two-way layout. CSSM B **6**, 49–61

980. ODEH, R. E. (1977): Extended tables of the distributions of rank statistics for treatment versus control in randomized block designs. CSSM B **6**, 103–113

981. ODEH, R. E. (1982): Critical values of the sample product-moment correlation coefficient in the bivariate distribution. CSSM **11**, 1–26

982. ODEH, R. E. (1982): Tables of percentage points of the distribution of the maximum absolute value of equally correlated normal random variables. CSSM **11**, 65–87

983. ODEH, R. E. and D. B. OWEN (1980): Tables for Normal Tolerance Limits, Sampling Plans, and Screening. (M. Dekker; pp. 316) New York

984. ODEH, R. E. and D. B. OWEN (1983): Attribute Sampling Plans, Tables of Tests and Confidence Limits for Proportions. (M. Dekker; pp. 392) New York

985. O'DONOVAN, T. M. (Ed; 1983): Short Term Forecasting: An Introduction to the Box Jenkins Approach. (Wiley; pp. 282) New York

986. OEHR, P. (1984): Tumormarker – Untersuchungen zum diagnostischen und prognostischen Wert. Medizinische Welt **35**, 1504–1512

987. ÖKSOY, D. and L. A. AROIAN (1982): Tables of the Distribution of the Correlation Coefficient. (Inst. of Administr. and Management, Union College; pp. 132) Schenectady, N. Y.

988. ÖKSOY, D. and L. A. AROIAN (1982): Percentage Points of the Distribution of the Correlation Coefficient. (Inst. of Administr. and Management, Union College; pp. 128) Schenectady, N. Y.

989. OJIMA, Y. (1983): Robustness of four standard methods for estimating the mean. JQTE **15**, 89–93

990. OMAN, S. D. (1984): Analyzing residuals in calibration problems. TECS **26**, 347–353

991. O'MUIRCHEARTAIGH, C. and D. P. FRANCIS (1981): Statistics: a dictionary of terms and ideas. (Arrow Book, The Anchor Press; pp. 295) Tiptree, Essex

992. O'MUIRCHEARTAIGH, C. A. and C. PAYNE (Eds.; 1977): The Analysis of Survey Data. I. Exploring Data Structures. II. Model Fitting. (Wiley; pp. 273, 255) New York

993. ONUKOGU, I. B. (1984): An analysis of variance of nominal data. Statistica **44**, 87–96

994. O'QUIGLEY, J. (1982): Regression models and survival prediction. STAN **31**, 107–116

995. ORD, J. K., G. P. PATIL and C. TAILLIE (Eds.; 1979): Statistical Distributions in Ecological Work. (Internat. Co-operative Publ. House; pp. 464) Fairland, Md.

996. ORLOCI, L. and N. C. KENKEL (1985): Introduction To Data Analysis. (International Co-operative Publ. House; pp. 340) Burtonsville, MD

997. ORLOCI, L., C. R. RAO and W. M. STITELER (Eds.; 1979): Multivariate Methods in Ecological Work. (International Co-operative Publ. House; pp. 550) Burtonsville, MD

998. ORTH, B. (1975): Einführung in die Theorie des Messens. (Kohlhammer; 132 S.) Stuttgart

999. OSAKI, S. and Y. HATOYAMA (Eds.; 1984): Stochastic Models in Reliability Theory. Proc. Symp. Nagoya, Japan April 23–24, 1984 (Lecture Notes in Economics and Mathematical Systems, Vol. 235) (Springer; pp. 212) Heidelberg and New York

1000. OTIS, D. L., K. P. BURNHAM, G. C. WHITE and D. R. ANDERSON (1978): Statistical Inference from Capture Data on Closed Animal Populations. (Wildlife Monographs, No. 62) (The Wildlife Society; pp. 135) Office of Academic Affairs, University of Louisville, Kentucky 40208 and Washington D. C. [Corrections to Wildlife Monograph 62, J. Wildl. Manage. **44** (3) (1980), 666–667]

1001. OTT, L. (1984): An Introduction to Statistical Methods and Data Analysis. 2nd ed. (Duxbury Press; pp. 775) Boston, Mass.

1002. OVERALL, J. E. (1980): Power of chi-square tests for 2×2 contingency tables with small expected frequencies. PYBU **87**, 132–135

1003. OVERALL, J. E. (1980): Continuity correction for Fisher's exact probability test. EDUC **5**, 177–190 [cf. 351–362]

1004. OVERALL, J. E. and R. R. STARBUCK (1983): F-test alternatives to Fisher's exakt test and to the chi-square test of homogeneity in 2×2 tables. EDUC **8**, 59–73

1005. OWEN, D. B. (1968): A survey of properties and applications of the noncentral t-distribution. TECS **10**, 445–478

1006. OWEN, D. B. (Ed.; 1976): On the History of Statistics and Probability. (Proc. Symp. on the American Mathematical Heritage, Southern Methodist Univ., Dallas, Tex. 1974) (M. Dekker; pp. 468) New York, Basel

1007. OWEN, D. B. and Loretta LI (1980): The use of cutting scores in selection procedures. EDUC **5**, 157–168

1008. OWEN, D. B., Loretta LI and Y.-M. CHOU (1981): Prediction intervals for screening using a measured correlated variate. TECS **23**, 165–170

1009. OWEN, G. (1982): Game Theory. 2nd ed. (Academic Press; pp. 368) New York

1010. OWEN, W. J. and T. A. DEROUEN (1980): Estimation of the mean for lognormal data containing zeroes and left-censored values, with applications to the measurement of worker exposure to air contaminants. BICS **36**, 707–719

1011. PACKEL, E. W. (1981): The Mathematics of Games and Gambling. (The Mathematical Association of America; Wiley; pp. 141) Washington

1012. PAGANO, M. and Katherine TAYLOR HALVORSEN (1981): An algorithm for finding the exact significance levels of $r \times c$ contingency tables. JASA **76**, 931–934

1013. PAGE, B. (1983): Der Gültigkeitsnachweis von komplexen Simulationsmodellen. Angewandte Informatik **25**, 149–157

1014. PAGE, E. (1982): Tables of waiting times for $M/M/n$, $M/D/n$ and $D/M/N$ and their use to give approximate waiting times in more general queues. J. Operational Res. Soc. **33**, 453–473

1015. Pagel, M. D. and C. E. Lunneborg (1985): Empirical evaluation of ridge regression. PYBU **97**, 342–355

1016. Pagnoni, A. and G. Rozenberg (Eds.; 1983): Application and Theory of Petri Nets. (Informatik-Fachberichte, Vol. 66) (Springer, pp. 315) Berlin, Heidelberg, New York, Tokyo

1017. Pahnke, K. (1984): Repeated Measures Modelle (RMM). Auswertung mit SAS und BMDP. SSNL **10**, 33–35

1018. Palachek, A. D. and W. R. Schucany (1983): On the correlation of a group of rankings with an external ordering relative to the internal concordance. Statistics and Probability Letters, (Amsterdam) **1**, 259–263

1019. Pandit, S. M. and S.-M. Wu (1983): Time Series and System Analysis with Applications. (Wiley; pp. 586) New York

1020. Pankratz, A. (1983): Forecasting with Univariate Box-Jenkins Models: Concepts and Cases. (Wiley; pp. 562) New York

1021. Park, C. N. and R. D. Snee (1983): Quantitative risk assessment: state-of-the-art for carcinogenesis. AMST **37**, 427–441

1022. Parker, R. G. and R. L. Rardin (1983): The travelling salesman problem: an update of research. Naval Research Logistics Quarterly **30**, 69–96

1023. Parr, W. C. and W. R. Schucany (1980): The jackknife: a bibliography. INTR **48**, 73–78

1024. Parzen, E. (1979): Non-parametric statistical data modeling. JASA **74**, 105–121 (and 121–131)

1025. Parzen, E. (1982): Data Modeling Using Quantile and Density-Quantile Functions. In: Tiago di Oliveira, J. and B. Epstein (Eds.): Some Recent Advances in Statistics. (Academic Press; pp. 248) London and New York, pp. 23–52

1026. Passing, H. (1984): Exact simultaneous comparisons with control in a $r \times c$ contingency table. BIJL **26**, 643–654

1027. Patefield, W. M. (1982): Exact tests for trends in ordered contingency tables. APST **31**, 32–43

1028. Patel, J. K., C. H. Kapadia and D. B. Owen (1976): Handbook of Statistical Distributions. (M. Dekker; pp. 302) New York, Basel

1029. Patel, J. K. and C. B. Read (1982): Handbook of the Normal Distribution (M. Dekker; pp. 337) New York and Basel

1030. Patil, G. P., M. T. Boswell, S. W. Joshi, M. V. Ratnaparkhi and J. J. J. Roux (1985): Dictionary and Classified Bibliography of Statistical Distributions in Scientific Work. Vol. 1. Discrete Models, Vol. 2. Univariate Continuous Models, Vol. 3. Multivariate Models. (International Co-operative Publ. House; pp. 458, 575, 350) Burtonsville, MD

1031. Patil, G. P. and M. L. Rosenzweig (Eds.; 1979): Contemporary Quantitative Ecology and Related Ecometrics. (International Co-operative Publ. House; pp. 695) Burtonsville, MD

1032. Patnaik, K. B. (1949): The non-central χ^2- and F-distributions and their applications. BIKA **36**, 202–232

1033. PATZAK, G. (1982): Systemtechnik – Planung komplexer innovativer Systeme. Grundlagen, Methoden, Techniken. (Springer; 445 S.) Berlin, Heidelberg, New York

1034. PAYNE, J. A. (1982): Introduction to Simulation. Programming Techniques and Methods of Analysis. (McGraw-Hill; pp. 324) New York

1035. PAYNE, W. H. (1977): Normal random numbers: Using machine language to choose the best algorithm. ACM Transactions on Mathematical Software **3**, 346–348

1036. PEARCE, S. C. (1983): The Agricultural Field Experiment. A Statistical Examination of Theory and Practice. (Wiley; pp. 335) New York

1037. PEARSON, E. S. (1978): The History of Statistics in the 17th and 18th Centuries. (Griffin; pp. 711) London

1038. PEARSON, E. S. and H. O. HARTLEY (Eds; 1970, 1972): Biometrika Tables for Statisticians. Vol. I and II (At the University Press; pp. 270 and pp. 385) Cambridge

1039. PEARSON, E. S., M. KENDALL and R. L. PLACKETT (Eds.; 1970, 1977): Studies in the History of Statistics and Probability. Vol. I and II (Griffin; pp. 491, 582) London and High Wycombe

1040. PEARSON, E. S. and N. W. PLEASE (1975): Relations between the shape of population distribution and the robustness of four simple test statistics. BIKA **62**, 223–241

1041. PEIL, J. und E. PESCHKE (1984): Methodische Aspekte der Aufbereitung von Meßwerten tierexperimenteller Stoffwechseluntersuchungen. Gegenbaurs morph. Jahrb. **130**, 531–555

1042. PEIL, J. und S. SCHMERLING (1984): Ein neues Verfahren zur Schätzung der Verteilung aus Meßwerten – dargestellt an morphometrischen Beispielen. Gegenbaurs morph. Jahrb. **130**, 973–800 (vgl. auch 779–792)

1043. PENDLETON, O. J. (1985): Influential observations in the analysis of variance. CSTH **14**, 551–565

1044. PENFIELD, D. A. and S. L. KOFFLER (1978): Post hoc procedures for some k-sample nonparametric tests for scale. EDUC **3**, 265–282

1045. PENTICO, D. W. (1981): On the determination and use of optimal sample sizes for estimating the difference in means. AMST **35**, 40–42

1046. PEREIRA, B. DE B. (1977): Discriminating among separate models: a bibliography. INTR **45**, 163–172

1047. PETERMANN, F. (1978): Veränderungsmessung. (Kohlhammer; 131 S.) Stuttgart [insbes. Kap. 9: Einzelfallbetrachtung]

1048. PETERSON, A. V. and L. D. FISHER (1980): Teaching the principles of clinical trials design and management. BICS **36**, 687–697

1049. PETTIT, A. N., and M. A. STEPHENS (1977): The Kolmogorov-Smirnov goodness-of-fit statistics with discrete and grouped data. TECS **19**, 205–210

1050. PFEFFERMANN, D. and T. M. F. SMITH (1985): Regression models for grouped populations in cross-section surveys. INTR **53**, 37–59

1051. PFEIFER, Ch. G. and P. ENIS (1978): Dorfman-type group testing for a modified binomial model. JASA **73**, 588–592

1052. PFOHL, H.-Ch. (1985): Logistiksysteme. Betriebswirtschaftliche Grundlagen. (Springer; 250 S.) Berlin, Heidelberg, New York, Tokyo

1053. PFOHL, H.-Ch. und G. E. BRAUN (1981): Entscheidungstheorie. Normative und deskriptive Grundlagen des Entscheidens. (Vlg. Moderne Industrie; 516 S.) Landsberg am Lech

1054. PILZ, L. und P. TAUTU (1983): Mathematische Modelle und ihre in numero Experimente in der Biologie. In C. O. Köhler u. a. (Hrsg.): Aktuelle Methoden der Information in der Medizin. Festschr. z. 65. Geburtstag v. Prof. Dr. med. Gustav Wagner. (ecomed; 243 S.) Landsberg/Lech, S. 83–109

1055. PITMAN, E. J. G. (1979): Some Basic Theory for Statistical Inference. (Chapman and Hall; pp. 110) London

1056. PLACKETT, R. L. (1981): Analysis of Categorical Data. 2nd rev. ed. (Griffin; pp. 215) London

1057. PLANE, D. R. and K. R. GORDON (1982): A simple proof of the non-applicability of the Central Limit Theorem to finite populations. AMST **36**, 175 + 176

1058. POCOCK, S. J. (1983): Clinical Trials. A Practical Approach. (Wiley; pp. 265) Chichester and New York

1059. POCOCK, S. J. and other conference participants (1982): Clinical trials. STAN **31**, 1–142

1060. POCOCK, S. J., Sheila M. GORE and Gillian R. KERR (1982): Long term survival analysis: the curability of breast cancer. SMED **1**, 93–104

1061. POKROPP, F. (1980): Stichproben – Theorie und Verfahren. (Athenäum-Vlg.; 255 S.) Königstein

1062. POLLARD, G. H. (1983): An analysis of classical and tie-breaker tennis. Austral. J. Statist. **25**, 496–505

1063. POLLARD, R. (1985): Goal-scoring and the negative binomial distribution. The Mathematical Gazette **69**, 45–47

1064. POLLOCK, K. H. (1981): Capure-recapture models allowing for age-dependent survival and capture rates. BICS **37**, 521–529

1065. POOLE, R. W. (1974): An Introduction to Quantitative Ecology. (McGraw-Hill; pp. 532) New York and Düsseldorf

1066. POPPER, K. R. (1982): Duldsamkeit und intellektuelle Verantwortlichkeit. In: Offene Gesellschaft – offenes Universum. Franz Kreuzer im Gespräch mit Karl R. Popper. (F. Deuticke; 118 S.) Wien, S. 103–116

1067. POPPER SHAFFER, Juliet (1977): Multiple comparisons emphasizing selected contrasts: an extension and generalization of Dunnett's procedure. BICS **33**, 293–303

1068. POPPER SHAFFER, Juliet (1981): Complexity: an interpretability criterion for multiple comparisons. JASA **76**, 395–401

1069. POTTER, R. and G. W. STURM (1981): The power of Jonckheere's test. AMST **35**, 249–250

1070. Potthoff, G. (1981): Statistische Klasseneinteilung. Wissenschaftl. Ztschr. d. Hochschule für Verkehrswesen "Friedrich List" Dresden **28**, 1049-1054

1071. Prakasa Rao, B.L.S. (1983): Nonparametric Functional Estimation. (Academic Press; pp. 522) Orlando, Florida

1072. Pratt, J. and J.D. Gibbons (1981): Concepts of Nonparametric Theory. (Springer; pp. 480) New York

1073. Preece, D.A. (1980): Covariance analysis, factorial experiments and marginality. STAN **29**, 97-122

1074. Preece, D.A. (1981): Distributions of final digits in data. STAN **30**, 31-60

1075. Preece, D.A. (1982): t is for trouble (and textbooks): a critique of some examples of the paired-samples t-test. STAN **31**, 169-195

1076. Preece, D.A. (1982): The design and analysis of experiments: what has gone wrong? Utilitas Mathematica **21**A, 201-244

1077. Pregibon, D. (1981): Logistic regression diagnostics. Annals of Statistics **9**, 705-724

1078. Press, S.J. (1982): Applied Multivariate Analysis: Using Bayesian and Frequentist Methods of Inference. 2nd ed. (R.E. Krieger; pp. 600) Florida

1079. Presse- und Informationsamt der Bundesregierung (Hrsg.; 1982): Gesellschaftliche Daten 1982. Reihe: Berichte und Dokumentation. 356 S., Bonn

1080. Priestley, M.B. (1981): Spectral Analysis and Time Series. Univariate Series. (Academic Press; pp. 736) London

1081. Profos, P. (1984): Meßfehler. Eine Einführung in die Meßtheorie. (Teubner; 140 S.) Stuttgart

1082. Prunty, L. (1983): Curve fitting with smooth functions that are piecewise-linear in the limit. BICS **39**, 857-866

1083. Quednau, H.D. (1983): Anwendungsmöglichkeiten der automatischen Mustererkennung in der Biologie. Medizinische Welt **34**, 659-665

1084. Raktoe, B.L., A. Hedayat and W.T. Federer (1981): Factorial Design. (Wiley; pp. 250) New York

1085. Ralston, A., H.S. Wilf and K. Enslein (Eds.; 1960, 1967, 1977): Mathematical Methods for Digital Computers. Vol I-III. (Wiley; pp. 293, 287, 454) New York

1086. Ramig, Pauline F. (1983): Applications of the analysis of means. JQTE **15**, 19-25

1087. Ramirez, M.M. (1984): A modification to some proposed tests in relation to the problem of switching regression models. CSTH **13**, 901-914

1088. Ramsay, P.H. (1981): Power of univariate pairwise multiple comparison procedures. PYBU **90**, 352-366

1089. Ramsay, P.H. (1982): Empirical power of procedures for comparing two groups on p variables. EDUC **7**, 139-156

1090. RANDLES, R. H. and D. A. WOLFE (1979): Introduction to the Theory of Nonparametric Statistics. (Wiley; pp. 450) New York

1091. RAO, B. R. and P. E. ENTERLINE (1984): Interaction contrast disease rates for assessing synergism (or antagonism) in multifactor-multilevel disease risks. BIJL **26**, 699–715

1092. RAPOPORT, A. (1980): Mathematische Methoden in den Sozialwissenschaften. (Physica-Verlag; 377 S.) Würzburg und Wien

1093. RASCH, D. (1978, 1984): Einführung in die Mathematische Statistik. 2. Aufl. I. Wahrscheinlichkeitsrechnung und Grundlagen der mathematischen Statistik. II. Regressionsanalyse, Varianzanalyse und weitere Anwendungen. (VEB Dtsch. Vlg. d. Wissenschaften; 372 S., 400 S.) Berlin

1094. RASCH, D. (Hauptautor; 1983): Biometrie. Einführung in die Biostatistik. (VEB Deutscher Landwirtschaftsverlag; 276 S.) Berlin

1095. RASCH, D., G. ENDERLEIN und G. HERRENDÖRFER (1973): Biometrie. Verfahren, Tabellen, angewandte Statistik. (VEB Dtsch. Landwirtschaftsverlag; 390 S.) Berlin

1096. RASCH, D., G. HERRENDÖRFER, J. BOCK und K. BUSCH (1978): Verfahrensbibliothek, Versuchsplanung und Auswertung. Bd. 1 und 2. (VEB Deutscher Landwirtschaftsverlag; 1052 S.) Berlin

1097. RASCH, D. and M. L. TIKU (Eds.; 1985): Robustness of Statistical Methods and Nonparametric Statistics. (D. Reidel; pp. 232) Dordrecht

1098. RATKOWSKY, D. A. (1983): Non-Linear Regression Modelling: A Unified Practical Approach. (M. Dekker; pp. 276) New York

1099. RAUHUT, B., N. SCHMITZ und W.-W. ZACHOW (1979): Spieltheorie. Eine Einführung in die mathematische Theorie strategischer Spiele. (Teubner; 400 S.) Stuttgart

1100. READ, C. B. (1977): Partitioning chi-square in contingency tables: A teaching approach. CSTH A **6**, 553–562

1101. READ, C. B. (1978): Test of symmetry in three-way contingency tables. PYKA **43**, 409–420

1102. REHM, N. (1976): Die Ermittlung des privaten Verbrauchs. Ein neuer Beitrag zur Fehlertheorie. (F. Steiner; 149 S.) Wiesbaden

1103. REHPENNING, W. (1983): Multivariate Datenbeurteilung. Statistische Untersuchungen über krankheitsbedingte Lage- und Strukturveränderungen klinisch-chemischer Kenngrößen. (Mediz. Informatik Statist., Bd. 43) (Springer, 89 S.) Berlin, Heidelberg, New York

1104. REISIG, W. (1982): Petrinetze. Eine Einführung. (Springer; 158 S.) Berlin, Heidelberg, New York

1105. REISIG, W. (1984): Petri Nets. An Introduction. (EATCS Monogr. on Theoret. Comp. Sci., Vol. 4) (Springer; pp. 161) Heidelberg and New York

1106. RENDALL, F. J. and D. M. WOLF (1983): Statistical Sources and Techniques. (McGraw-Hill; pp. 245) London

1107. RETZLAFF, G., G. RUST und J. WAIBEL (1978): Statistische Versuchspla-

nung. Planung naturwissenschaftlicher Experimente und ihre Auswertung mit statistischen Methoden. 2. verb. Aufl. (Verlag Chemie; 211 S.) Weinheim 1978

1108. REUTER, A. und U. REIMER (1984): Knowledge Base Management System. Informatik-Spektrum 7, 44–45 und 173

1109. REVENSTORF, D. (1980): Faktorenanalyse. (Kohlhammer; 188 S.) Stuttgart

1110. REYNOLDS, H. T. (1977): The Analysis of Cross-Classifications. (The Free Press; pp. 236) New York

1111. RHOADES, H. M. and J. E. OVERALL (1982): A sample size correction for Pearson chi-square in 2×2 contingency tables. PYBU 91, 418–423

1112. RICHARDS, W. D., Jr. (1980): Simulation. Chapter 15 in P. R. Monge and J. N. Cappella (Eds.): Multivariate Techniques in Human Communication Research. (Academic Press; pp. 552) New York and London, pp. 455–487

1113. RICHARDSON, J. (Ed.; 1984): Models of Reality: Shaping Thought and Action. (Lomond; in conjunction with UNESCO; pp. 328) Mt. Airy, Maryland

1114. RIEDWYL, H. (1980): Graphische Gestaltung von Zahlenmaterial. 2. Aufl. UTB 4440 (P. Haupt; 167 S.) Bern und Stuttgart

1115. RINDSKOPF, D. (1984): Linear equality restrictions in regression and loglinear models. PYBU 96, 597–603

1116. RINGLAND, J. T. (1983): Robust multiple comparisons. JASA 78, 145–151

1117. RIPLEY, B. D. (1983): Computer generation of random variables: a tutorial. INTR 51, 301–319

1118. RIPLEY, B. D. (1984): Spatial statistics: developments 1980–3. INTR 52, 141–150

1119. RIVETT, P. (1980): Model Building for Decision Analysis. (Wiley; pp. 172) New York

1120. ROBERTS, F. S. (1979): Measurement Theory. With Applications to Decisionmaking, Utility and the Social Sciences. (Addison-Wesley; pp. 420) Reading, Mass.

1121. ROBERTS, H. V. and R. F. LING (1983): Conversational Statistics with IDA. (McGraw-Hill; pp. 666) New York

1122. ROCHEL, H. (1983): Planung und Auswertung von Untersuchungen im Rahmen des allgemeinen linearen Modells. (LFT Psychologie, Bd. 4) (Springer; 262 S.) Berlin, Heidelberg, New York

1123. ROCKE, D. M. (1983): Robust statistical analysis of interlaboratory studies. BIKA 70, 421–431

1124. ROCKHOLD, F. W. and S. J. KILPATRICK (1981): Methods of discrimination among stochastic models of the negative binomial distribution with an application to medical statistics. BIJL 23, 681–692

1125. RODGERS, W. L. (1984): An evaluation of statistical matching. Journal of Business and Economic Statistics 2, 91–102

1126. ROGER, J. H. (1977): A significance test for cyclic trends in incidence data. BIKA **64**, 152–155

1127. ROGERS, W. H. and J. W. TUKEY (1972): Understanding some longtailed distributions. STNE **26**, 211–226

1128. ROGOSA, D. (1980): Comparing nonparallel regression lines. PYBU **88**, 307–321

1129. ROMESBURG, H. Ch. (1984): Cluster Analysis for Researchers. (Lifetime Learning Publications; pp. 334) Belmont, Calif. 94002

1130. ROSEMANN, H. (1981): Zuverlässigkeit und Verfügbarkeit technischer Anlagen und Geräte. Mit praktischen Beispielen von Berechnung und Einsatz in Schwachstellenanalysen. (Springer; 188 S.) Heidelberg

1131. ROSENBAUM, P. R. and D. B. RUBIN (1985): Constructing a control group using multivariate matched sampling methods that incorporate the propensity score. AMST **39**, 33–38

1132. ROSENTHAL, R. (1984): Meta-Analytic Procedures for Social Research. (Sage: Appl. Soc. Res. Meth. Ser., Vol. 6; pp. 160) Beverly Hills and London

1133. ROSENTHAL, R. and D. B. RUBIN (1983): Ensemble-adjusted p values. PYBU **94**, 540–541 [see also **97** (1985), 521–529]

1134. ROSNER, B. (1982): A generalization of the paired t-test. APST **31**, 9–13

1135. ROSNER, B. (1983): Percentage points for a generalized ESD many-outlier procedure. TECS **25**, 165–172

1136. ROSS, S. M. (1982): Stochastic Processes. (Wiley; pp. 320) New York

1137. ROSSI, P. H., J. D. WRIGHT and A. B. ANDERSON (Eds.; 1983): Handbook of Survey Research. (Academic Press; pp. 800) New York and London

1138. ROTHERY, P. (1979): A nonparametric measure of intraclass correlation. BIKA **66**, 629–639

1139. ROTHMAN, K. J. (1976): Causes. AMEP **104**, 587–592

1140. ROTHMAN, K. J. and J. D. BOICE, Jr. (1982): Epidemiologic Analysis with a Programmable Calculator. 2nd ed. (The New England Epidemiology Institute; pp. 197) Chestnut Hill, PO Box 57, MA 02167

1141. ROYEN, Th. (1984): Multiple comparisons of polynomial distributions. BIJL **26**, 319–332

1142. RUBIN, D. B., T. W. F. STROUD and Dorothy T. THAYER (1981): Fitting additive models to unbalanced two-way data. EDUC **6**, 153–178

1143. RUBINSTEIN, R. Y. (1981): Simulation and the Monte Carlo Method. (Wiley; pp. 300) New York

1144. RUDICH, A. and W. GERISCH (1984): MOCAFICO – A Monte Carlo computer program for the critical values of Fisher's test on the significance in harmonic analysis and of Cochran's test on the homogeneity of variances. Computational Statistics Quarterly (Vienna) **1**, 77–84

1145. RÜMKE, Chr. L. (1982): How long is the wait for an uncommon event? CHRO **35**, 561–564

1146. RÜPPEL, H. (1977): Bayes-Statistik. Eine Alternative zur klassischen Statistik. Archiv für Psychologie **129**, 175–186

1147. RÜTZEL, E. (1979): Bayessches Hypothesentesten und warum Bayesianer Bias-ianer heißen sollten. Archiv für Psychologie **131**, 211–232

1148. RUPP, S. und K. SCHWARZ (Hrsg.; 1983): Beiträge aus der bevölkerungswissenschaftlichen Forschung: Festschrift für Herrmann Schubness. (H. Boldt-Verlag; 592 S.) Boppard am Rhein

1149. RUSTAGI, J. S. and D. A. WOLFE (Eds.; 1982): Teaching of Statistics and Statistical Consulting. (Academic Press; pp. 548) New York

1150. SAATY, Th. L. and J. M. ALEXANDER (1981): Thinking With Models. Mathematical Models in the Physical, Biological and Social Sciences. (Pergamon; pp. 181) Oxford and New York

1151. SACHS, L. (1982): Statistische Methoden. 5. neubearb. Aufl. (Springer; 124 S.) Berlin Heidelberg, New York

1152. SACHS, L. (1983): Angewandte Statistik. Anwendung Statistischer Methoden. 6. neubearb. Aufl. (Springer; 552 S.) Berlin, Heidelberg, New York

1153. SACHS, L. (1984): Applied Statistics. A Handbook of Techniques. 2nd ed. Translated from the German by Z. Reynarowych. (Springer Series in Statistics). (Springer; pp. 707) New York, Berlin, Heidelberg, Tokyo

1154. SÄRNDAL, C. E. (1974): A comparative study of association measures. PYKA **39**, 165–187

1155. SAGER, T. W. (1983): Estimating modes and isopleths. CSTH **12**, 529–557

1156. SAHAI, H. (1979): A bibliography on variance components. INTR **47**, 177–222

1157. SAHAI, H., A. I. KHURI, and C. H. KAPADIA (1985): A second bibliography on variance components. CSTH **14**, 63–115

1158. SAKAROVITCH, M. (1983): Linear Programming. (Springer; pp. 210) Berlin, Heidelberg, New York

1159. SAMPFORD, M. R. (1962): An Introduction to Sampling Theory. With Applications to Agriculture. (Oliver and Boyd; pp. 292) Edinburgh and London

1160. SAMPSON, A. R. and R. L. SMITH (1982): Assessing risks through the determination of rare event probabilities. Operations Research **30**, 839–866

1161. SANDE, I. G. (1982): Imputation in surveys: coping with reality. AMST **36**, 145–152

1162. SANDVIK, L. and Birgitta OLSSON (1982): A nearly distribution-free test for comparing dispersion in paired samples. BIKA **69**, 484–485

1163. SAVORY, S. E. (Hrsg.; 1985): Künstliche Intelligenz und Expertensysteme. Ein Forschungsbericht der Nixdorf Computer AG. (R. Oldenbourg; 248 S; [Kap. 1 in engl. Spr.]) München und Wien

1164. SAW, J. G., M. C. K. YANG and T. C. MO (1984): Chebyshev inequality with estimated mean and variance. AMST **38**, 130–132

1165. SAYERS, B. (1981): The risk assessment of large technological systems. With discussion. Bull. Int. Statist. Inst. **49**, 444–461

1166. SCHACH, Elisabeth und S. SCHACH (1978): Pseudoauswahlverfahren bei Personengesamtheiten I: Namensstichproben. ASTA **62**, 379–396

1167. SCHACH, Elisabeth und S. SCHACH (1979): Pseudoauswahlverfahren bei Personengesamtheiten II: Geburtstagsstichproben. ASTA **63**, 108–122

1168. SCHAEFER, E. (1979): Zuverlässigkeit, Verfügbarkeit und Sicherheit in der Elektronik. (Vogel; 368 S. [182 Abb., 2farbig]) Würzburg

1169. SCHAEFER, F. (ADM, Hrsg.; 1979): Muster – Stichproben – Pläne. (Vlg. Moderne Industrie; 160 S.) München

1170. SCHAFER, W. D. (1980): Assessment of dispersion in categorical data. Educational und Psychological Measurement **40**, 179–183

1171. SCHAFFRANEK, M. (1980): Wirtschafts- und Bevölkerungsstatistik. (Kohlhammer; 206 S.) Stuttgart

1172. SCHAICH, E. (1983): Die Anwendung von Stichproben bei Inventuren. ASTA **67**, 274–285

1173. SCHARF, J.-H. (1981): Möglichkeiten der mathematischen Formulierung von Wachstumsprozessen. Gegenbaurs morph. Jahrb. **127**, 706–740

1174. SCHEAFFER, R. L. (1980): Multiple comparisons for Poisson rates. JQTE **12**, 94–97

1175. SCHECHTMAN, Edna (1982): A nonparametric test for detecting changes in location. CSTH **11**, 1475–1482

1176. SCHEFFÉ, H. (1973): A statistical theory of calibration. Annals of Statistics **1**, 1–37

1177. SCHEMPER, M. (1981): Spezielle Verfahren und Programme zur statistischen Analyse zensierter Daten. EDVM **12**, 42–45 [vgl. auch 23–26]

1178. SCHEMPER, M. (1985): Statistical methods and programs for nonparametric analysis of pairs. SSNL **11**, 128–129

1179. SCHERRER, G. und I. OBERMEIER (1981): Stichprobeninventur. Theoretische Grundlagen und praktische Anwendung. (Vahlen; 178 S.) München

1180. SCHILLER, Karla und E. SONNEMANN (1981): Tests zum multiplen Niveau. Bemerkungen und Ergänzungen zu den HOLM-Prozeduren. Arbeitsber. d. Abtlg. Statistik d. Univ. Dortmund Nr. 10 (Mai 1981), 40 S.

1181. SCHILLING, E. G. (1982): Acceptance Sampling in Quality Control. (M. Dekker; pp. 775) New York

1182. SCHLESSELMAN, J.J. (1982): Case Control Studies. Design, Conduct, Analysis. (Oxford Univ. Press; pp. 354) New York

1183. SCHLITTGEN, R. (1979): Use of a median test for a generalized Behrens-Fisher problem. MEKA **26**, 95–103

1184. SCHLITTGEN, R. und B. H. J. STREITBERG (1984): Zeitreihenanalyse. (R. Oldenbourg; 476 S.) München und Wien

1185. SCHLÖRER, J. (1982): Outputkontrollen zur Sicherung statistischer Datenbanken. Informatik-Spektrum **5**, 224–236

1186. SCHMEISER, B. W. (1980): Generation of variates from distribution tails. Operations Research **28**, 1012–1017

1187. SCHMERLING, S., J. RENNE, W. LENZE und G. BÄTZ (1981): Probleme der Erweiterung der Prüfverteilungen des Maximum-Modulus-Testes und des Dunnett-Testes. BIJL **23**. 29-40

1188. SCHMID, C. F. (1983): Statistical Graphics: Design, Principles and Practices. (Wiley; pp. 224) New York

1189. SCHMID, C. F. and S. E. SCHMID (1979): Handbook of Graphic Presentation. 2nd ed. (Ronald Press, Wiley; pp. 308) New York

1190. SCHMITZ, N. und F. LEHMANN (1976): Monte-Carlo Methoden I: Erzeugen und Testen von Zufallszahlen. (A. Hain; 134 S.) Meisenheim am Glan

1191. SCHNEEBERGER, H. (1971): Optimierung in der Stichprobentheorie durch Schichtung und Aufteilung. Unternehmensforschung **15**, 240-254

1192. SCHNEIDER, B. und U. RANFT (1978): Simulationsmethoden in der Medizin und Biologie. (Med. Inf. Stat., Bd. 8) (Springer; 496 S.) Heidelberg

1193. SCHÖNE, A. (1981): Prozeßrechnersysteme. Aufbau und Programmierung von Prozeßrechnern. Grundlagen und Verfahren ihrer Anwendung. (Hanser; 714 S.) München und Wien

1194. SCHREGENSBERGER, J. W. (1982): Methodenbewußtes Problemlösen. Ein Beitrag zur Ausbildung von Konstrukteuren, Beratern und Führungskräften. (P. Haupt; 232 S.) Bern, Stuttgart

1195. SCHUCHARD-FICHER, C., K. BACKHAUS, U. HUMME, W. LOHRBERG, W. PLINKE und W. SCHREINER (1982): Multivariate Analysenmethode. Eine Anwendungsorientierte Einführung. 2. Aufl. (Springer; 346 S.) Heidelberg [3. Aufl., 1985]

1196. SCHULZINGER-FOX, Jane (1984): SIR Primer. (SIR, STATUS GmbH; pp. 120) D-1000 Berlin 45 and Skokie, Illinois 60077

1197. SCHUMACHER, J. und R. VOLLMER (1981): Partnerwahl und Partnerbeziehung. Die Gravitation des Partnermarktes und ihre demographischen Folgen. Zeitschrift für Bevölkerungswissenschaft **7**, 499-518

1198. SCHUMACHER, M. (1981): Power and sample size determination in survival time studies with special regard to the censoring mechanism. MIME **20**, 110-115

1199. SCHUMACHER, R. B. (1981): Systematic measurement error. JQTE **13**, 10-24

1200. SCHUMAN, H. and S. PRESSER (1981): Questions and Answers in Attitude Surveys. Experiments on Question Form, Wording, and Context. (Academic Press; pp. 392) New York and London

1201. SCHWARZ, H. (1975): Stichprobenverfahren. Ein Leitfaden zur Anwendung statistischer Schätzverfahren. (Oldenbourg; 194 S.) München und Wien

1202. SCHWARZ, K. (1982): Bericht 1982 über die demographische Lage in der Bundesrepublik Deutschland. Zeitschrift für Bevölkerungswissenschaft **8**, 121-223

1203. SCHWARZE, J. und J. WECKERLE (Hrsg.; 1982): Prognoseverfahren im

Vergleich. Anwendungserfahrungen und Anwendungsprobleme verschiedener Prognoseverfahren. (Prof. Dr. J. Schwarze, TU, Postf. 3329; 220 S.) Braunschweig

1204. SCHWING, R. C. and W. R. ALBERS, Jr. (Eds.; 1980): Societal Risk Assessment. How Safe is Safe Enough? (Plenum Press; pp. 372) New York

1205. SCOTT, D. T., G. R. BRYCE and D. M. ALLEN (1985): Orthogonalization-triangularization methods in statistical computation. AMST **39**, 128–135

1206. SCOTT, D. W. (1979): On optimal and data-based histograms. BIKA **66**, 605–610

1207. SEARLE, S. R. (1982): Matrix Algebra Useful for Statistics. (Wiley; pp. 438) New York

1208. SEBER, G. A. F. (1982): Estimation of Animal Abundance and Related Parameters. 2nd rev. ed. (Griffin; pp. 600) London

1209. SEBER, G. A. F. (1984): Multivariate Observations. (Wiley; pp. 686) New York

1210. SEN, A. R. (1983): Review of some important techniques in wildlife sampling and sampling errors. BIJL **25**, 699–715

1211. SEN, I. and V. P. PRABHASHANKER (1980): A nomogram for estimating the three parameters of the Weibull distribution. JQTE **12**, 138–143

1212. SHABAN, S. A. (1980): Change point problem and two-phase regression: an annotated bibliography. INTR **48**, 83–93

1213. SHAH, B. V. (1984): Software for survey data analysis. AMST **38**, 68–69

1214. SHAPIRO, S. H. and T. A. LOUIS (Eds.; 1983): Clinical Trials: Issues and Approaches. (M. Dekker; pp. 209) New York

1215. SHAPIRO, S. S. and C. W. BRAIN (1982): Recommended distributional testing procedures. Amer. J. Math. Manag. Sci. **2**, 175–221

1216. SHERIF, Y. S. (1982): Reliability analysis: optimal inspection and maintenance schedules for failing systems. Microelectron. Reliab. **22**, 59–115

1217. SHERIF, Y. S. and M. L. SMITH (1981): Optimal maintenance models for systems subject to failure – a review. Naval Research Logistics Quarterly **28**, 47–74

1218. SHEYNIN, O. B. (1983): Corrections and short notes on my papers. Archive for History of Exact Sciences **28**, 171–195

1219. SHIGEMASU, K. (1982): A Bayesian analysis of covariance from comparing changes. BEHA **12**, 85–96

1220. SHIRAHATA, S. (1982): Nonparametric measures of intraclass correlation. CSTH **11**, 1707–1721

1221. SHIRAHATA, S. (1982): A nonparametric measure of interclass correlation. CSTH **11**, 1723–1732

1222. SHIUE, W.-K. and L. J. BAIN (1983): A two-sample test of equal gamma distribution scale parameters with unknown common shape parameter. TECS **25**, 377–381

1223. SHUMWAY, R. and J. GURLAND (1960): Fitting the Poisson binomial distribution. BICS **16**, 522–533

1224. Silver, M. S. (1982): On an awareness of alternative approaches to index numbers. Teaching Statistics **4**, 80–84

1225. Simar, L. (1984): A survey of Bayesian approaches to nonparametric statistics. Math. Operationsforsch. u. Statist., series statistics **15**, 121–142

1226. Singpurwalla, N. D. and M.-Y. Wong (1983): Estimation of the failure rate: A survey of nonparametric methods. Part I: Non-Bayesian methods. CSTH **12**, 559–588

1227. Skilling, J. H. (1980): On the null distribution of Jonckheere's statistic used in two-way models for ordered alternatives. TECS **22**, 431–436

1228. Skillings, J. H. (1983): Nonparametric approaches to testing and multiple comparisons in a one-way ANOVA. CSSM **12**, 373–387

1229. Skillings, J. H. and G. A. Mack (1981): On the use of a Friedman type statistic in balanced and unbalanced block designs. TECS **23**, 171–177

1230. Slome, C., Donna Brogan, Sandra Eyres and W. Lednar (1982): Basic Epidemiological Methods and Biostatistics: A Workbook. (Wadsworth; pp. 354) Belmont, Calif.

1231. Slovic, P., B. Fischhoff and Sarah Lichtenstein (1979): Rating the risk. Environment **21**, (1979), 14–20 and 36–39

1232. Smith, D. and N. Keyfitz (1977): Mathematical Demography. Selected Papers. (Springer; pp. 514) New York, Heidelberg, Berlin

1233. Smith, J. H. and T. O. Lewis (1982): Effects of intraclass correlation on covariance analysis. CSTH **11**, 71–80

1234. Smith, J. R. and J. M. Beverly (1981): The use and analysis of staggered nested factorial designs. JQTE **13**, 166–173

1235. Smith, Patricia (1979): Splines as a useful and convenient statistical tool. AMST **33**, 57–62

1236. Smith, P. J. and S. C. Choi (1982): Simple tests to compare two dependent regression lines. TECS **24**, 123–126

1237. Smith, T. M. F. (1983): On the validity of inferences from nonrandom samples. JRSA **146**, 394–403

1238. Snedecor, G. W. and W. G. Cochran (1980): Statistical Methods. 7nth ed. (Iowa State Univ. Press; pp. 507) Ames, Iowa

1239. Snee, R. D. (1983): Graphical analysis of process variation studies. JQTE **15**, 76–88

1240. Solberg, H. E. (1978): Discriminant analysis. Critical Reviews in Clinical Laboratory Sciences **9**, 209–242

1241. Solomon, H. (1982): Measurement and Burden of Evidence. In: Tiago di Oliveira, J. and B. Epstein (Eds.): Some Recent Advances in Statistics. (Academic Press; pp. 248) London and New York, pp. 1–22

1242. Sonnemann, E. (1982): Allgemeine Lösungen multipler Testprobleme. EDVM **13**, 120–128

1243. Sonquist, J. A. and W. C. Dunkelberg (1977): Survey and Opinion Research: Procedures for Processing and Analysis. (Prentice-International; pp. 502) Englewood Cliffs, N. J.

1244. Soper, J. B. (1983): Helping students to select the appropriate formula. Teaching Statistics **5**, 20–24

1245. Späth, H. (1983): Cluster-Formation und -Analyse: Theorie, FORTRAN Programme, Beispiele. (R.Oldenbourg; 236 S.) München und Wien

1246. Spector, P. E. (1981): Research Designs. (Sage Publicat. Series 07-023; pp.80) Beverly Hills and London

1247. Spjøtvoll, E. and A. H. Aastveit (1980): Comparison of robust estimators on data from field experiments. Scand. J. Stat. **7**, 1–13

1248. Spjøtvoll, E. and A. H. Aastveit (1983): Robust estimators on laboratory measurements of fat and protein in milk. BIJL **25**, 627–639

1249. SPSS-X User's Guide (1983): (McGraw-Hill; pp.806) New York

1250. SPSS-X Basics (1984): (McGraw-Hill; pp.214) New York

1251. Spurrier, J. D., J. E. Hewett and Z. Lababidi (1982): Comparison of two regression lines over a finite interval. BICS **38**, 827–836

1252. Stange, K. (1977): Bayes-Verfahren. (Springer; 312 S.) Berlin, Heidelberg, New York

1253. Stanish, W. M. and N. Taylor (1983): Estimation of the intraclass correlation coefficient for the analysis of covariance model. AMST **37**, 221–224

1254. Starr, T. B. (1983): A method for calculating a confidence interval for the ratio of two Poisson parameters. AMEP **118**, 785

1255. Statistisches Bundesamt (Hrsg.; 1984): Statistisches Jahrbuch 1984 für die Bundesrepublik Deutschland. (W. Kohlhammer; 790 S.) Stuttgart und Mainz [Stat. Jb. 1985, 776 S., 1985]

1256. Steiger, J. H. (1980): Tests for comparing elements of a correlation matrix. PYBU **87**, 245–251

1257. Steinberg, D. M. and W. G. Hunter (1984): Experimental design: review and comment. With discussion. TECS **26**, 71–130

1258. Steinhausen, D. und K. Langer (1977): Clusteranalyse. Einführung in Methoden und Verfahren der automatischen Klassifikation. (de Gruyter; 206 S.) Berlin und New York

1259. Steinhorst, R. K. (1982): Resolving current controversies in analysis of variance. AMST **36**, 138–139

1260. Stem, D. E. and R. K. Steinhorst (1985): Telephone interview and mail questionnaire applications of the randomized response model. JASA **79**, 555–564

1261. Stenger, H. (1971): Stichprobentheorie. (Physica-Vlg.; 228 S.) Würzburg und Wien

1262. Stephan, F. F. und P. J. McCarthy (1958): Sampling Opinions. An Analysis of Survey Procedure. (Wiley; pp.451) New York

1263. Sternberg, R. J. (1977): Writing the Psychology Paper (Barron's Educat. Series; pp.243) Woodbury, N. Y.

1264. Stevens, J. P. (1984): Outliers and influential data points in regression analysis. PYBU **95**, 334–344

1265. STIER, W. (1980): Verfahren zur Analyse saisonaler Schwankungen in ökonomischen Zeitreihen. (Springer; 134 S.) Heidelberg

1266. STIRLING, W. D. (1982): Enhancements to aid interpretation of probability plots. STAN **31**, 211–220

1267. STÖRMER, H. (1983): Mathematische Theorie der Zuverlässigkeit. Einführung und Anwendungen. (Oldenbourg: 329 S.) München und Wien

1268. STOLINE, M. R. (1981): The status of multiple comparisons: simultaneous estimation of all pairwise comparisons in one-way ANOVA designs. AMST **35**, 134–141

1269. STOLINE, M. R. and H. K. URY (1979): Tables of the studentized maximum modulus distribution and an application to multiple comparisons among means. TECS **21**, 87–93

1270. STRUBE, M. J. (1985): Combining and comparing significance levels from nonindependent hypothesis tests. PYBU **97**, 334–341 [for more on meta-analysis see 251–306]

1271. STUART, A. (1984): The Ideas of Sampling. (Griffin; pp. 91) London

1272. STURGEON, M. L. (1980): Describing the response surface in exploratory two-drug chemotherapy experiments. Computers and Biomedical Research **13**, 1–18

1273. SUBRAHMANIAM, K. and A. V. GAJJAR (1980): Robustness to nonnormality of some transformations of the sample correlation coefficient. Journal of Multivariate Analysis **10**, 60–77

1274. SUDMAN, S. (1976): Applied Sampling. (Academic Press; pp. 249) New York and London

1275. SUDMAN, S. (1980): Reducing response error in surveys. STAN **29**, 237–273

1276. SUDMAN, S. and N. M. BRADBURN (1983): Asking Questions. A Practical Guide to Questionnaire Design. (Jossey-Bass, SAGE Publ.; pp. 384) London

1277. SUSSER, M. and Zena STEIN (1982): Third variable analysis: application to causal sequences among nutrient intake, maternal weight, birthweight, placental weight, and gestation. SMED **1**, 105–120

1278. SUZUKI, M. (1983): Estimation in a bivariate semi-lognormal distribution. BEHA **13**, 59–68

1279. SWAFFORD, M. (1980): Three parametric techniques for contingency table analysis: a nontechnical commentary. American Sociological Review **45**, 664–690

1280. SWALLOW, W. H. and J. R. TROUT (1983): Determination of limits for a linear regression or calibration curve. JQTE **15**, 118–125

1281. SWAMURTHY, M. (1982): Growth and Structure of Human Population in the Presence of Migration. Studies in Population. (Academic Press; pp. 227) New York

1282. SYRBE, M. (1984): Zuverlässigkeit von Realzeitsystemen: Fehlermanagement. Informatik-Spektrum **7**, 94–101

1283. Szameitat, K. (1982): Zur Analyse und Weiterverarbeitung des Datenmaterials in der amtlichen Statistik. Angew. Stat. Ökonom. **21**, 193–212

1284. Tabachnik, Barbara G. and Linda S. Fidell (1983): Using Multivariate Statistics. (Harper & Row; pp.509) New York

1285. Tadikamalla, P. R. (1980): A look at the Burr and related distributions. INTR **48**, 337–344

1286. Tallarida, R. J. and R. B. Murray (1981): Manual of Pharmacological Calculations with Computer Programs. (Springer; pp.150) New York, Heidelberg, Berlin

1287. Talwar, P. P. and J. E. Gentle (1981): Detecting a scale shift in a random sequence at an unknown time point. APST **30**, 301–304

1288. Tamhane, A. C. (1979): A comparison of procedures for multiple comparisons of means with unequal variances. JASA **74**, 471–480

1289. Tapia, R. A. and J. R. Thompson (1978): Nonparametric Probability Density Estimation. (John Hopkins Univ. Press; pp.176) Baltimore

1290. Tarone, R. E. (1979): Testing the goodness of fit of the binomial distribution. BIKA **66**, 585–590

1291. Tarone, R. E. (1981): On the distribution of the maximum of the logrank statistic and the modified Wilcoxon statistic. BICS **37**, 79–85

1292. Tarone, R. E. (1981): On summary estimators of relative risk. CHRO **34**, 463–468

1293. Taub, T. W. (1979): Computation of two-tailed Fisher's test. JQTE **11**, 44–47

1294. Thakur, J. K. (1984): A FORTRAN program to perform the nonparametric Terpstra-Jonckheere test. Computer Programs in Biomedicine **18**, 235–240

1295. Theobald, C. M. and R. L. Mallinson (1978): Comparative calibration, linear structural relationship and congeneric measurements. BICS **34**, 39–45

1296. Thissen, D., L. Baker and H. Wainer (1981): Influence-enhanced scatter plots. PYBU **90**, 179–184

1297. Thöni, H. (1977): Testing the difference between two coefficients of correlation. BIJL **19**, 355–359

1298. Tholey, P. (1981/82): Signifikanztest und Bayessche Hypothesenprüfung. Archiv für Psychologie **134**, 319–342

1299. Thomas, D. G. and J. J. Gart (1977): A table of exact confidence limits for differences and ratios of two proportions and their odds ratios. JASA **72**, 73–76

1300. Thomas, H. (1982): Decision Analysis in Insurance. (Croom Helm; pp.192) London

1301. Thomas, H. (1982): Decision Analysis in the Pharmaceutical Industry. (Croom Helm; pp.192) London

1302. Thomas, L. C. (1985): Games, Theory and Applications. (Horwood; pp.279) Chichester

1303. THOMPSON, B. (1985): Canonical Correlation Analysis. (Sage Publicat. Series 07-047; pp. 69) Beverly Hills and London

1304. THOMPSON, W. A., Jr. (1981): On the foundations of reliability. TECS **23**, 1-13

1305. TIDMORE, F. E. and D. W. TURNER (1983): On clustering with Chernoff-type faces. CSTH **12**, 381-396

1306. TIETJEN, G. L. and M. E. JOHNSON (1979): Exact statistical tolerance limits for sample variances. TECS **21**, 107-110

1307. TIKU, M. L. (1982): Testing linear contrasts of means in experimental design without assuming normality and homogeneity of variances. BIJL **24**, 613-627

1308. TILLMAN, F. A., C.-L. HWANG and W. KUO (1980): Optimization of Systems Reliability. (M. Dekker; pp. 311) New York

1309. TILLMAN, F. A., W. KUO, C.-L. HWANG and D. L. GROSH (1982): Baysian reliability and availability. A review. IEEE Trans. Reliab. R-**31**, 362-372

1310. TÖRNQVIST, L., P. VARTIA and Y. O. VARTIA (1985): How should relative changes be measured? AMST **39**, 43-46

1311. TÖWE, J., J. BOCK and G. KUNDT (1985): Interactions in contingency table analysis. BIJL **27**, 17-24

1312. TRAUBOTH, H. und A. JAESCHKE (Hrsg.; 1984): Prozeßrechner 1984. Prozeßdatenverarbeitung im Wandel, 4. GI/GMR/GFK-Fachtagung Karlsruhe, 26.-28. Sept. 1984. (Informatik-Fachberichte, Bd. 86) (Springer; 710 S.) Berlin, Heidelberg, New York

1313. TRAUT, H. (1983): Significance tests in mutagen screening: another method considering historical control frequencies. BIJL **25**, 717-719

1314. TRUETT, Jeanne, J. CORNFIELD and W. KANNEL (1967): A multivariate analysis of the risk of coronary heart disease in Framingham. J Chronic Diseases **20**, 511-524 [see also **27** (1974), 97-102]

1315. TUCKER, A. (1980): Applied Combinatorics. (Wiley:. pp. 385) New York

1316. TUCKER, R. K. and L. J. CHASE (1980): Canonical Correlation. Chapter 7 in P. R. Monge and J. N. Cappella (Eds.): Multivariate Techniques in Human Communication Research. (Academic Press; pp. 552) New York and London, pp. 205-228

1317. TUFTE, E. R. (1983): The Visual Display of Quantitative Information. (Graphics Press; pp. 190) Chesshire, CT

1318. TUKEY, J. W. (1969): Analyzing data. Sanctification of detective work? American Psychologist **24**, 83-91

1319. TUKEY, J. W. (1972): Some Graphic and Semigraphic Displays. In: Statistical Papers in Honor of George W. Snedecor, ed. by T. A. Bancroft and Susan A. Brown, Chapter 18, pp. 293/316. The Iowa Univ. Press; Ames, Iowa

1320. TUKEY, J. W. (1977): Exploratory Data Analysis. (Addison-Wesley; pp. 688) Reading, Mass.

1321. TUKEY, J. W. (1979): Methodology and the statistician's responsibility for BOTH accuracy AND relevance. JASA **74**, 786-793

1322. TUKEY, J. W. (1980): We need both exploratory and confirmatory. AMST **34**, 23-25

1323. TUKEY, J. W. (1980): Methodological Comments Focused on Opportunities. Chapter 16 in P. R. Monge and J. N. Cappella (Eds.): Multivariate Techniques in Human Communication Research. (Academic Press; pp. 552) New York and London, pp. 489-528

1324. TURIEL, T. O., G. J. HAHN and W. T. TUCKER (1982): New simulation results for the calibration and inverse median estimation problems. CSSM **11**, 677-713

1325. TYEBJEE, T. T. (1979): Telephone survey methods. The state of the art. Journal of Marketing **43**, 68-78

1326. TYGSTRUP, N., J. M. LACHIN and E. JUHL (Eds.; 1982): The Randomized Clinical Trial and Therapeutic Decisions. (M. Dekker; pp. 320) New York

1327. TZAFESTAS, S. G. (1980): Optimization of system reliability: a survey of problems and techniques. Int. J. Systems Sci. **11**, 455-486

1328. UEHLINGER, H.-M. (1983): Datenverarbeitung und Datenanalyse mit SAS. Eine problemorientierte Einführung. (G. Fischer Vlg.; 322 S.) Stuttgart und New York

1329. UHLMANN, W. (1982): Statistische Qualitätskontrolle. Eine Einführung. 2. überarb. u. erweit. Aufl. (Teubner; 292 S.) Stuttgart

1330. ULLMAN, J. D. (1983): Principles of Database Systems. 2nd ed. (Pitman; pp. 484) London

1331. University Microfilms International (1984): Comprehensive Dissertation Index 1983 Supplement, Vol. 2: Sciences, Part 2 (UMI; pp. 498) Section: Mathematics and Statistics, pp. 329-384. UMI, 300 North Zeeb Road, P. O. Box 1764, Ann Arbor, Mich. 48106

1332. UNKELBACH, H. D. und T. WOLF (1985): Qualitative Dosis-Wirkungs-Analysen. Einzelsubstanzen und Kombinationen. (G. Fischer; 137 S.) Stuttgart und New York

1333. UNWIN, D. (1981): Introductory Spatial Analysis. (Methuen; pp. 212) London

1334. UPTON, G. J. G. (1978): The Analysis of Cross-tabulated Data. (Wiley; pp. 148) New York

1335. UPTON, G. J. G. (1982): A comparison of alternative tests for the 2×2 comparative trial. JRSA **145**, 86-105

1336. URY, H. K. (1977): A comparison of some approximations to the Wilcoxon-Mann-Whitney distribution. CSSM B **6**, 181-197

1337. URY, H. K. (1982): Comparing two proportions; finding p_2 when p_1, n, α and β are specified. STAN **31**, 245-250

1338. VAN ES, A. J., R. D. GILL and C. VAN PUTTEN (1983): Random number generators for a pocket calculator. STNE **37**, 95-102

1339. VELLEMAN, P. F. and D. C. HOAGLIN (1981): Applications, Basics and Computing of Exploratory Data Analysis. (Duxbury Press; pp. 354) Boston, Mass.

1340. VELLEMAN, P. F. and R. E. WELSCH (1981): Efficient computing of regression diagnostics. AMST **35**, 234–242

1341. VEN, VAN DER, A. (1980): Einführung in die Skalierung. (Übers. u. herausgegeb. v. J. Groebel) (Huber; 409 S.) Bern, Stuttgart, Wien

1342. VERMA, V. (1981): Assessment of errors in household surveys. Bull. Int. Statist. Inst. **49**, 905–919

1343. VERMA, V. (1982): The Estimation and Presentation of Sampling Errors. (Internat. Statist. Institute; pp. 59) Voorburg/NL

1344. VIANA, M. A. (1980): Statistical methods for summarizing independent correlational results. EDUC **5**, 83–104

1345. VICTOR, N. (1978): Alternativen zum klassischen Histogramm. MIME **17**, 120–126

1346. VICTOR, N. (1984): Computational statistics. Tool or science? – Werkzeug oder Wissenschaft? With discussion. SSNL **10**, 105–125

1347. VICTOR, N., W. LEHMACHER und W. VAN EIMEREN (Hrsg.; 1980): Explorative Datenanalyse. (Mediz. Informatik und Statist. Bd. 26) (Springer; 221 S.) Berlin, Heidelberg, New York

1348. VICTOR, N., J. DUDECK und E. P. BROSZIO (Hrsg.; 1981): Therapiestudien. (Mediz. Informatik Statist., Bd. 33) (Springer; 600 S.) Berlin, Heidelberg, New York

1349. VINEK, G., P. F. RENNERT und A. M. TJOA (1982): Datenmodellierung: Theorie und Praxis des Datenbankentwurfs. (Physica-Vlg.; 306 S.) Würzburg und Wien

1350. VINOD, H. D. and A. ULLAH (1981): Recent Advances in Regression Methods. (M. Dekker; pp. 361) New York

1351. VOGEL, F. (1980): Was ist und was kann die Statistik? Wirtschaftswissenschaftliches Studium **9**, 288–293

1352. VOGEL, F. (1984): Beschreibende und schließende Statistik. Formeln, Definitionen, Erläuterungen, Stichwörter und Tabellen. 2. Aufl. (R. Oldenbourg; 196 S.) München und Wien

1353. VOLLMAR, J. (Hrsg.; 1983): Biometrie in der chemisch-pharmazeutischen Industrie I. (G. Fischer; 162 S.) Stuttgart und New York

1354. WACHOLDER, S. and C. R. WEINBERG (1982): Paired versus two-sample design for a clinical trial of treatments with dichotomous outcome: power considerations. BICS **38**, 801–812

1355. WAGENER, C. (1984): Diagnostic sensitivity, diagnostic specificity and predictive value of the determination of tumor markers. Journal of Clinical Chemistry and Clinical Biochemistry **22**, 969–979

1356. WAGNER, G. und E. GRUNDMANN (Hrsg.; 1983): Basisdokumentation für Tumorkranke. Prinzipien und Verschlüsselungsanweisungen für Klinik und Praxis im Auftrag der Arbeitsgemeinschaft Deutscher Tumorzentren (ADT). 3. erw. Aufl. (Springer; 85 S.) Berlin, Heidelberg, New York

1357. WAHRENDORF, J. (1980): Inference in contingency tables with ordered categories using Plackett's coefficient of association for bivariate distributions. BIKA **67**, 15–21

1358. WAINER, H. (1976): Robust statistics: a survey and some prescriptions. EDUC **1**, 285–312

1359. WAINER, H. and D. THISSEN (1981): Graphical Data Analysis. Annual Review of Psychology **32**, 191–241

1360. WALKER, S. H. and D. B. DUNCAN (1967): Estimation of the probability of an event as a function of several independent variables. BIKA **54**, 167–179

1361. WALLACE, W. L. (1983): Principles of Scientific Sociology. (Aldine; pp. 545) New York

1362. WALLIS, W. A. (1942): Compounding probabilities from independent significance tests. Econometrica **10**, 229–248

1363. WALTER, S. D. (1976): The estimation and interpretation of attributable risk in health research. BICS **32**, 829–849

1364. WALTER, S. D. (1978): Calculation of attributable risks from epidemiological data. Internat. J. Epidemiol. **7**, 175–182 [cf. also AMEP **122** (1985), 904–914]

1365. WAMPOLD, B. E. (1984): Tests of dominance in sequential categorical data. PYBU **96**, 424–429

1366. WANG, P. C. (Ed.; 1978): Graphical Presentation of Multivariate Data. (Academic Press; pp. 288) New York

1367. WANG, P. P. (Ed.; 1983): Advances in Fuzzy Sets, Possibility Theory, and Applications. (Plenum Publ. Corp.; pp. 421) New York

1368. WARE, J. H. (1985): Linear models for the analysis of longitudinal studies. AMST **39**, 95–101

1369. WARREN, W. G. (1971): Correlation or regression: Bias or precision. APST **20**, 148–164

1370. WARREN, W. G. (1982): On the adequacy of the chi-squared approximation for the coefficient of variation. CSSM **11**, 659–666

1371. Warwick, D. P. and C. A. LININGER (1975): The Sample Survey: Theory and Practice. (McGraw-Hill; pp. 344) New York and London

1372. WASSERMAN, P. and J. O'BRIEN (Eds.; 1982): Statistics Sources. A Subject Guide to Data on Industrial, Business, Social, Educational, Financial, and other Topics for the United States and Internationally. (Gale Research; pp. 1388) Detroit

1373. WATSON, G. S. (1983): Statistics on Spheres. (Wiley; pp. 238) New York

1374. WEBER, Erna (1980): Grundriß der biologischen Statistik. 8. überarb. Aufl. (G. Fischer; 652 S.) Stuttgart

1375. WEBER, K., R. TREBZINER und H. TEMPELMEIER (1983): Simulation mit GPSS. (P. Haupt; 489 S.) Bern–Stuttgart

1376. WEBER, W. E. und H. P. LIEBIG (1981): Anpassung einer Ausgleichsfunktion an beobachtete Werte. EDVM **12**, 88–92

1377. WEED, R. M. (1982): Bounds for correlated compound probabilities. JQTE **14**, 196–200

1378. WEGMAN, E. J., El-Sayad NOUR and C. KUKUK (1980): A time series approach to life table construction. CSTH A **9**, 1587–1607

1379. WEGMAN, E.J. and I.W. WRIGHT (1983): Splines in statistics. JASA **78**, 351–365

1380. WEINBERG, R. and Y.C. PATEL (1981): Simulated intraclass correlation coefficients and their z transforms. JSCS **13**, 13–26

1381. WEINBERG, S.L. and R.B. DARLINGTON (1976): Canonical analysis when the number of variables is large relative to sample size. EDUC **1**, 313–332

1382. WEISBERG, H.F. and B.D. BOWEN (1977): An Introduction to Survey Research and Data Analysis. (W.H. Freeman; pp. 243) Reading, England and San Francisco, USA

1383. WEISBERG, S. (1980): Applied Linear Regression (Wiley; pp. 283) New York

1384. WERMUTH, Nanny (1978): Zusammenhangsanalysen Medizinischer Daten. (Mediz. Informatik u. Statist., Bd. 5) (Springer; 115 S.) Berlin, Heidelberg, New York

1385. WERNER, J. und B. WERNER (1984): Ridge-Regression: kein Routine-Verfahren. PYBE **26**, 283–297, 537

1386. WERTZ, W. (1978): Statistical Density Estimation: A Survey. (Vandenhoeck and Ruprecht; pp. 108) Göttingen

1387. WESTGARD, J.O., R.N. CAREY and S. WOLD (1974): Criteria for judging precision and accuracy in method development and evaluation. Clinical Chemistry **20**, 825–833

1388. WETHERILL, G.B. (1975): Sequential Methods in Statistics. 2nd ed. (Chapman and Hall; pp. 242) London

1389. WETHERILL, G.B. (1977): Sampling Inspection and Quality Control. 2nd ed. (Chapman and Hall; pp. 154) London

1390. WETHERILL, G.B. (1981): Intermediate Statistical Methods. (Chapman and Hall; pp. 390) London [Solutions to Exercises, pp. 74, 1981]

1391. WHITE, G.C., D.R. ANDERSON, K.P. BURNHAM and D.L. OTIS (1982): Capture–Recapture and Removal Methods for Sampling Closed Populations. (LA-8787-NERP; UC-11) (National Laboratory; pp. 235) Los Alamos, NM 87545

1392. WHITE, J.S. (1970): Tables of the normal percentile points. JASA **65**, 635–638

1393. WHITEHEAD, J. (1983): The Design and Analysis of Sequential Clinical Trials. (Wiley; pp. 272) New York

1394. WIDDRA, W. (1972): Eine Verallgemeinerung des "Gesetzes seltener Ereignisse". MEKA **19**, 68–71

1395. WILBURN, A.J. (1984): Practical Statistical Sampling for Auditors. (M. Dekker; pp. 410) New York

1396. WILCOX, R.R. (1981): A review of the beta-binomial model and its extensions. EDUC **6**, 3–32

1397. WILCOX, R.R. (1983): A table of percentage points of the range of independent t variables. TECS **25**, 201–204

1398. WILDT, A. R. and O. T. AHTOLA (1978): Analysis of Covariance. (Sage Publicat. Series 07-012; pp. 91) Beverly Hills and London

1399. WILKE, J. (1978): Bibliographie. Multivariate Statistik und Mehrdimensionale Klassifikation. Bd. 1 und 2 (Akademie Vlg.; 1122 S.) Berlin

1400. WILKIE, D. (1981): Birthdays and breakdowns revisited. Teaching Statistics (Sheffield) **3**, 17-21

1401. WILL, U. (1985): Induktion und Rechtfertigung. (V. Klostermann; 210 S.) Frankfurt am Main

1402. WILLAN, A. R. and D. G. WATTS (1978): Meaningful multicollinearity measures. TECS **20**, 407-412

1403. WILLIAMS, B. (1978): A Sampler on Sampling. (Wiley; pp. 254) New York

1404. WILLIGAN, J. D. and Katherine A. LYNCH (1982): Sources and Methods of Historical Demography. (Academic Press; pp. 505) New York and London

1405. WILSON, R. and E. A. C. CROUCH (1982): Risk/Benefit Analysis. (Ballinger; pp. 218) Cambridge, Mass.

1406. WILSON, S. R. (1979): Examination of regression residuals. Austral. J. Statist. **21**, 18-29

1407. WILSON, S. R. (1983): Benchmark data sets for the flexible evaluation of statistical software. COMP **1**, 29-39

1408. WINER, B. J. (1971): Statistical Principles in Experimental Design. 2nd ed. (McGraw-Hill; pp. 907) New York

1409. WINKLER, W. (1969): Darstellung und Messung der Bevölkerungsverteilung im geographischen Raum. MEKA **14**, 138-163

1410. WINKLER, W. (1983): Vorlesungen zur Mathematischen Statistik. (Teubner; 276 S.) Stuttgart

1411. WIRTH, E. (1979): Theoretische Geographie. Grundzüge einer Theoretischen Kulturgeographie. (Teubner; 335 S.) Stuttgart

1412. WISHART, D. (1984): CLUSTAN Benutzerhandbuch (3. Ausgabe). (aus dem Engl. übers. von J. B. Schäffer) (G. Fischer, 244 S.) Stuttgart

1413. WITTE, E. H. (1980): Signifikanztest und statistische Inferenz. Analysen, Probleme, Alternativen. (Enke; 193 S.) Stuttgart

1414. WOELFEL, J. and J. E. DANES (1980): Multidimensional Scaling Models for Communication Research. Chapter 11 in P. R. Monge and J. N. Cappella (Eds.): Multivariate Techniques in Human Communication Research. (Academic Press; pp. 552) New York and London, pp. 333-364

1415. WOHLFAHRT, S. (1983): Wohnumfeldstruktur und Aktivitäten älterer Menschen. Zeitschrift für Bevölkerungswissenschaft **9**, 93-107

1416. WOLD, S. (1974): Spline functions in data analysis. TECS **16**, 1-11

1417. WOLF, E. H. and J. I. NAUS (1973): Tables of critical values for a k-sample Komolgorov-Smirnov test statistic. JASA **68**, 994-997

1418. WOLFE, D. A. (1977): A distribution-free test for related correlation coefficients. TECS **19**, 507-509

1419. WOLTER, K. M. (1985): Introduction to Variance Estimation. (Springer; pp. 427) New York, Berlin, Heidelberg, Tokyo

1420. WONNACOTT, T. H. and R. J. WONNACOTT (1981): Regression: A Second Course in Statistics. (Wiley; pp. 556) New York

1421. WOODBURY, M. A., K. G. MANTON and E. STALLARD (1981): Longitudinal models for chronic disease risk: an evaluation of logistic multiple regression and alternatives. International Journal of Epidemiology **10**, 187–197

1422. WOODING, W. O. (1973): The split plot design. JQTE **5**, 16–33

1423. WOODWARD, W. A. and A. C. ELLIOT (1983): (Microcomputer Software Reviews:) Observations on microcomputer programs for statistical analysis. SSNL **9**, 52–60

1424. WRIGHT, T. (Ed.; 1983): Statistical Methods and the Improvement of Data Quality. (Academic Press; pp. 384) New York and London

1425. WRIGLEY, N. (1979): Development in the statistical analysis of categorical data. Progress in Human Geography **3**, 315–355

1426. WT-GEIGY (1980): Wissenschaftliche Tabellen Geigy, Teilband Statistik, 1980, 8. rev. u. erweiterte Aufl. (der "Documenta Geigy") (CIBA-GEIGY AG; 241 S.) Basel

1427. WU, G. T., S. L. TWOMEY and R. E. THIERS (1975): Statistical evaluation of method-comparison data. Clinical Chemistry **21**, 315–320

1428. WUNSCH, G. J. and M. G. TERMOTE (1978): Introduction to Demographic Analysis. Principles and Methods. (Plenum Press; pp. 274) New York and London

1429. WYNDER, E. L. (Managing Ed.; 1982): Weak associations in epidemiology and their interpretation. Preventive Medicine **11**, 464–476

1430. XEKALAKI, Evdokia (1984): The bivariate generalized Waring distribution and its application to accident theory. JRSA **147**, 488–498

1431. YAMANE, T. (1967): Elementary Sampling Theory. (Prentice-Hall; pp. 405) Englewood Cliffs, N. J.

1432. YANG, M. C. K. and R. L. CARTER (1983): One-way analysis of variance with time series data. BICS **39**, 747–751

1433. YATES, F. (1981): Sampling Methods for Censuses and Surveys. 4. rev., enlarged ed. (Griffin; pp. 458) London

1434. YATES, F. (1984): Tests of significance for 2×2 contingency tables. With discussion. JRSA **147**, 426–463

1435. YOUDEN, W. J. (1972): Enduring values. TECS **14**, 1–11

1436. YOUNG, F. W. (1981): Quantitative analysis of qualitative data. PYKA **46**, 357–388

1437. YOUNGER, Mary S. (1979): A Handbook for Linear Regression. (Duxbury; pp. 569) North Scituate, Mass.

1438. ZACKS, S. (1982): Classical and Bayesian approaches to the change-point problem: fixed sample and sequential procedures. Statistique et Analyse des Données **7**, 48–81

207

1439. ZADEH, L. A. (1984): Coping with the imprecision of the real world. An interview. Communications of the ACM **27**, 304-311

1440. ZAHN, D. A. (1975): Modification of and revised critical values for the half-normal plot. TECS **17**, 189-200 [cf. also 201-211]

1441. ZAHN, D. A. and D. J. ISENBERG (1983): Nonstatistical aspects of statistical consulting. AMST **37**, 297-302

1442. ZAR, J. H. (1978): Approximations for the percentage points of the chi-square distribution. APST **27**, 280-290

1443. ZAR, J. H. (1984): Statistical significance of mutation frequencies, and the power of statistical testing, using the Poisson distribution. BIJL **26**, 83-88

1444. ZIELKE, W. (1980): Handbuch Lern-, Denk-, Arbeitstechniken. (W. Dummer; 438 S.) München 50

1445. ZIMMERMAN, H. J., B. R. GAINES and L. A. ZADEH (Eds.; 1984) Fuzzy Sets and Decision Analysis. (North-Holland; pp.504) Amsterdam, New York, Oxford

1446. ZIMMERMANN, H. (1984): Die praktische Relevanz des McNemar-Tests. BIJL **26**, 219-220

1447. ZIMMERMANN, H. (1984): Erweiterung des Mantel-Haenszel-Tests auf $K(>2)$ Gruppen. BIJL **26**, 223-224

1448. ZWICK, Rebecca and L. A. MARASCUILO (1984): Selection of pairwise multiple comparison procedures for parametric and nonparametric analysis of variance Models. PYBU **95**, 148-155

1449. ZWICK, Rebecca, Virginia NEUHOFF, L. A. MARASCUILO and J. R. LEVIN (1982): Statistical tests for correlated proportions: some extensions. PYBU **92**, 258-271

6. Software: Some Addresses/Einige Anschriften

AIDA
  Action-Research Northwest, 11442 Marine View Drive, S. W.
  Seattle, WA 98146, U.S.A.

BMDP
  BMDP Statistical Software, Inc., 1964 Westwood Blvd. Suite 202
  Los Angeles, CA 90025, U.S.A.

  [BMDP] Statistical Software Ltd., Cork Farm Centre, Dennehy's Cross
  Cork, Ireland

GLIM
  NAG Central Office, 256 Banbury Road
  Oxford, OX2 7 DE, United Kingdom

IMSL
  International Mathematical and Statistical Libraries, Inc.,
  7500 Bellaire Blvd.
  Houston, TX 77036, U.S.A.

LISREL
  Scientific Software Inc., P.O. Box 536
  Mooresville, IN 46158, U.S.A.

MARC
  Inter/View B.V., P.O. Box 60068
  1005 GB Amsterdam, The Netherlands

MINITAB
  Minitab Project, 215 Pond Laboratory
  University Park, PA 16802, U.S.A.

MSUSTAT
  Research and Development Institute, Inc., Montana State University
  Bozeman, MT 59717-0002, U.S.A.

NAG
   NAG Central Office, 256 Banbury Road
   Oxford, OX2 7 DE, United Kingdom

OSIRIS
   Institute for Social Research, University of Michigan, P.O. Box 1248
   Ann Arbor, MI 48106, U.S.A.

P-STAT, Inc.
   P.O. Box AH, 471 Wall Street
   Princeton, NJ 08540, U.S.A.

QSL/Bellview
   Questionnaire Specification, Language and Interviewing System,
   Pulse Train Technology, Ltd., 15 Lakeside Drive
   Esher, Surrey KT10 9EZ, United Kingdom

S
   Bell Laboratories, Computer Information Service, 600 Mountain Avenue
   Murray Hill, NJ 07974, U.S.A.

   Technology Licensing Manager, AT and T, P.O. Box 25000
   Greensboro, NC 27420, U.S.A.

SAS
   SAS Institute Inc., Box 8000, SAS Circle
   Cary, NC 27511-8000, U.S.A.

   SAS Institute Inc., PO Box 10066
   Raleigh, NC 27605, U.S.A.

   SAS Institute GmbH, Rohrbacher Str. 22
   D-6900 Heidelberg, F.R.G.

SIGSTAT
   Prof. Dr. R.C. Galbraith, Significant Statistics, 3336 North Canyon Road
   Provo, UT 84604, U.S.A.

SIR
   SIR Inc., 5215 Old Orchard Rd.
   Skokie, IL 60077, U.S.A.

STATUS GmbH
   Postfach 450149
   Goerzallee 5
   D-1000 Berlin 45, F.R.G.

SPSS
SPSS Inc., 444 North Michigan Ave
Chicago, IL 60611, U.S.A.

SPSS-Europe B.V.
Avelingen West 5, P.O. Box 115
N-4200 AC Gorinchem, The Netherlands

Mr. Major Lester/SPSS UK Ltd.
London House, 243-253 Lower Mortlake Rd.
Richmond, Surrey TW9 2LL, United Kingdom

STATCALC
Profs. Drs. A.J. Lee and P. McInerey, Dept. of Mathematics and Statistics,
University of Auckland
Auckland, New Zealand

STAT 80
Angela Hampton, Statware, P.O. Box 51 08 81
Salt Lake City, UT 84151, U.S.A.

STATGRAPHICS
Statistical Graphics Corporation, Research Park/2 Wall Street
Princeton, N.J. 08540, U.S.A.

STATPAK
Northwest Analytical Inc., P.O. Box 1 44 30
Portland, ORE 97214, U.S.A.

SURVO
Prof. Dr. S. Mustonen, University of Helsinki, Dept. of Statistics,
Alekanterinkatu 7
SF-00100 Helsinki, Finland

SYSTAT, Inc.
603 Main St.
Evanston, IL 60202, U.S.A.

TAU
Office of Population Censuses and Surveys, Titchfield, Fareham
Hants PO15 5RR, United Kingdom

The Consistent System
Renaissance Computing, Inc., P.O. Box 699
Cambridge, MA 02139, U.S.A.

TPL
U.S. Bureau of Labor Statistics, 441 G St. N.W.
Washington, D.C. 20212, U.S.A.

TSP
TSP International, 204 Junipero Serra Blvd.
Stanford, CA 94305, U.S.A.

L. Sachs

Applied Statistics

A Handbook of Techniques

Translated from the German by
Z. Reynarowych

2nd edition. 1984. 59 figures. XXVIII, 707 pages.
(Springer Series in Statistics). ISBN 3-540-90976-1

Contents: Introduction. – Introduction to Statistics. – Preliminaries. – Statistical Decision Techniques. – Statistical Methods in Medicine and Technology. – The Comparison of Independent Data Samples. – Further Test Procedures. – Measures of Association: Correlation and Regression. – The Analysis of $K \times 2$ and Other Two Way Tables. – Analysis of Variance Techniques. – Bibliography and General References. – Exercises. – Solutions to the Exercises. – Few Names and Some Page Numbers. – Subject Index.

Applied Statistics is an up-dated translation of a highly successful German book which has gone through five editions. Slanted toward and containing many illustrative examples from medicine and biology, its coverage of basic statistics – including a wide variety of techniques, tables, and computational aids – is notably thorough and complete. An unusual feature of this book is its comprehensive review of the literature on statistical tests, design of experiments, and other aspects of applied statistics.

Springer-Verlag
Berlin Heidelberg New York
London Paris Tokyo

Springer

Springer Series in Statistics

Advisors: **D. Brillinger, S. Fienberg, J. Gani, J. Hartigan, K. Krickeberg**

D. F. Andrews, A. M. Herzberg

Data:

A Collection of Problems from Many Fields for the Student and Research Worker

1985. 11 figures, 100 tables. XX, 442 pages.
ISBN 3-540-96125-9

This unique collection of statistical tables contains a wealth of empirical, theoretically relevant, and fruitful actual data. Statisticians will use it with benefit in their discussions of methods and approaches, and it is also an important tool for teachers and students who want to understand the relevance and applicability of specific statistical methods. Altogether, the volume contains more than 100 tables and graphs, containing thousands of interesting and illuminating data.

Springer-Verlag
Berlin Heidelberg New York
London Paris Tokyo

Springer

5-12-87